ALSO BY PAUL HAWKEN

Seven Tomorrows:
Toward a Voluntary History

The Next Economy

Growing a Business

The Ecology of Commerce: A
Declaration of Sustainability

Natural Capitalism: The
Next Industrial Revolution

Blessed Unrest: How the Largest
Movement in the World Came into Being,
and Why No One Saw It Coming

Drawdown: The Most Comprehensive
Plan Ever Proposed to Reverse
Global Warming (Editor)

Regeneration: Ending the Climate
Crisis in One Generation

CARBON

THE BOOK OF LIFE

Paul Hawken

VIKING

VIKING
An imprint of Penguin Random House LLC
1745 Broadway, New York, NY 10019
penguinrandomhouse.com

Grateful acknowledgment is made to Báyò Akómoláfé for permission
to reprint the excerpt on page 1 from "Foreword" on his website,
www.bayoakomolafe.net.

Designed by Alexis Farabaugh

LIBRARY OF CONGRESS CONTROL NUMBER: 2024036141
ISBN 9780525427445 (hardcover)
ISBN 9780698173293 (ebook)

Printed in the United States of America
1st Printing

The authorized representative in the EU for product safety and compliance is
Penguin Random House Ireland, Morrison Chambers, 32 Nassau Street,
Dublin D02 YH68, Ireland, https://eu-contact.penguin.ie.

I dedicate this work to the Traditional
Guardians of the land upon which it was written. Seven
generations of my family have benefited and been nurtured
by the animals, plants, and practices of the Coastal Miwok
nations. Their connection to the sea, rivers, forests, and
grasslands is a constant reminder and teaching. I offer
my respect to elders past, present, and emerging.

I also dedicate these writings to Jasmine
Scalesciani Hawken, whose love, kindness, and
untiring support uplifted and inspired me throughout.

CONTENTS

CARBON

ONE

Carbon

There are things we must do, sayings we must say, thoughts we must think, that look nothing like the images of success that have so thoroughly possessed our visions of justice.

BÁYÒ AKÓMOLÁFÉ

Carbon moves ceaselessly through the four realms—the biosphere, oceans, land, and atmosphere. It flows in rivers and veins, soil and skin, breath and wind. It is the narrator of lives born and lost, futures feared and imagined. It is the courier coursing through every particle of our existence, the interwoven lattice that permeates cultures, lagoons, minds, grasslands, organisms, and our temporal life. Carbon's dance of life does not take sides; it is never right or wrong. It is a timeless path that endlessly unwinds before us. Like Ariadne's thread, the flow of carbon is a story that may allow us to escape the labyrinth of anxiety, ignorance, and fear the world bequeaths. Carbon's increase in the atmosphere moves in tandem with the loss of the living world. *The Book of Life* encircles what has always regulated climate, the pulsing, living mantle we call Earth.

Like you, I take in the news, the science, the confusion, the broken politics—a world unfurled, fearful, at wit's end, shrouded in shallow certainties. To better understand the riddles and luminosity of life, I chose to go far upstream, to headwaters, and look at the flow of life through the lens of carbon. Rather than bemoan the plight of the world solely through forecasts and portents, I turned to voices who see the planet absent the overlay of threats. Might there be wisdom domes as well as heat domes? There are women and men merging observational Indigenous wisdom and Western science into a different understanding of our place on Earth, a perspective that reveals what we do not know. Certainties are dissolving. They are being replaced by unfathomable complexity. Though carbon comprises a tiny fraction of the Earth, a planet without it is a dead rock in space, like a sky without stars, a symphony without sound. We have reduced carbon to an errant element, the culprit in a civilization bent on self-termination. The crises of a warming planet, rampant injustice, and collapsing biodiversity form a whole. Carbon, people, and nature are set apart as if they were independent. Carbon is a window into the entirety of life, with all its beauty, secrets, and complexity. When discussing carbon, people refer to atoms instead of magnificence, physics rather than sentience. Life is a flow, a river, not isolated components. Stubborn beliefs, petty details, and irrelevant media can splinter our awareness. The flow of carbon provides better stories, other ways to see, visions of possibility different from the disjointed and chaotic narratives that engulf us.

From a planetary view, the warming atmosphere is a response, an adjustment, a teaching. Earth's climate is not breaking down as some would have it. However, it is changing faster than humans can adapt. Global heating foretells a tumultuous future. If human-induced greenhouse gas emissions are not curtailed, civilization will be. After decades of unwavering coaching by climate scientists, the world has awakened to climate dynamics. The changing atmosphere is front and center for companies, countries, schools, and universities. Investors are creating the most significant capital event in human history. Climate will be the fulcrum of finance for decades to come. Although banks, investors, and pension funds were once apathetic to financing a livable future, the prospect of decarbonizing the $110 trillion global economy has changed many minds. What is on the agenda? Every home, car, train, plane, truck, city, ship, product, farm, building, and utility in the world. Regarding resources, all wood, steel, concrete, fiber, plastics, and minerals.

For industry, the changing climate is seen as an engineering problem, not a crisis of behavior, consumption, or disconnection. There is a tacit assumption that the current fossil fuel-based energy system can be swapped out for renewables and the privileged can continue to live the way they do. This is magical thinking. To remedy global warming, oil companies strive to capture and remove carbon from the atmosphere as if it were an overflowing storage depot. It is emblematic of how business has come to perceive the Earth—a manageable contrivance humans can service, modify, and fix. It implies that a juggernaut economy can tame the atmosphere with claims of being carbon neutral. The current lifestyle of the world is maintained at the cost of a terrifying future. There is no defense for our misguided conduct and the disintegration of the living world.

Entrepreneurs have created carbon dioxide markets, as was once done with enslaved human beings and ivory tusks. There is now a marketplace for biodiversity credits. The International Monetary Fund calculated the value of a blue whale at $2 million—a so-called nature-based solution, a term that implies we can fix the natural world the same way we are attempting to repair the atmosphere. What could the monetization of a whale possibly mean? The unswerving belief in the marketplace as a means to create a better world is belied by history. Extracting and selling the biosphere to the highest bidder is the cause of global warming and social injustice. Stepping back from the inordinate obsession with wealth, it is apparent that commerce is eliminating life on Earth to pay shareholder dividends.

When Prince Hamlet lamented, "There's the rub," he was contemplating suicide and realized it required leaving his mortal coil. The rub for civilization is the curious, delusional beliefs of commerce. Citizen Potawatomi biologist Robin Wall Kimmerer explains the snag: "We need more than policy change; we need a change of worldview, from the fiction of human exceptionalism to the reality of our kinship and reciprocity with the living world. The planet asks us that we renounce a culture of endless taking so that the world can continue." This cannot happen if political, financial, and corporate powers think solely about future gains. The task of modernity is to recognize that our existence rests upon the entirety of planetary life.

The world economy is undergoing a massive energy transition; a civilization based on fossil-fuel combustion is transforming into one powered by current solar income: solar panels, wind turbines, and hydro. The necessity is clear. Governing and financial institutions required decades to embrace the climate crisis. Yet, now that they do, the dominant discourse about the crisis places the living world into a subordinate position, a distinct category, essential to be sure, but not as urgent, usually referred to as biodiversity. How greenhouse gases change the atmosphere is well known. How trillions of living entities regulate the atmosphere and generate the bounty of our home planet is not understood. Bioethicist Melanie Challenger describes how "we are trying to design life on our own terms even while we are killing life on its terms." As human wants continue to unravel the planet's capacity to regenerate, we enter an unimaginable future of biological poverty, where our attempt to redress the atmosphere will hardly matter. In all of Earth's multibillion-year history, that which did not work, that which did not serve life, was discarded. Why are we in that queue?

Millions of years of earthly wealth have been consumed and eliminated in the past two centuries. Reefs are perishing, pollinators are declining, oceans are acidifying, fisheries are ransacked, forests are toppling, soils are eroding, lands are desiccating, birds are vanishing, and wildlands are dwindling. A future can only be grasped if there is an accurate understanding of the present. We are attempting to sever the human world from the natural world as if that were possible. The current system of production and consumption eats its host. Enshrined economic practices beget and ensure the losses. Challenger writes, "Our cities and industries have left their imprint in the soil, in the cells of deep-sea creatures, in the distant particles of the atmosphere. The trouble is that we don't know the right way to behave towards life. This uncertainty exists in part because we can't decide how other life-forms matter or even if they do."

Replacing fossil fuels with renewables is crucial but insufficient. Humanity depends upon its relationship to all of Earth's habitats and denizens, even if we don't think so. Society, commerce, and governments must focus on what journalist Eric Roston calls the dance of carbon, the constant regeneration inherent to life. This does not preclude technical innovation and invention. Technologies are needed that pass an essential threshold: Does a solution, stratagem, or proposal create more life or less? We have tried less, and this is where it has brought us. What does more look like? Pure water, clean food, vibrant cultures, honored people, ancient forests, human health, equity, education, abundant fisheries, wildness, quiet green cities, rich soil, living wages, and dignified work.

Though largely ignored by the media and news feeds, the movement to regenerate the living world exists in thousands of organizations and millions of people. Life-giving communities are smaller, submerged, and unnoticed by mega-institutions whose marketing, publicity, and social media dominate our lives. The actions of citizen-led and Indigenous communities are based on reciprocity, mutualism, and reconciliation with the natural world, qualities that do not lend themselves to the news cycle. Their work reflects what evolutionary biologist Peter Kropotkin noted early in the last century: cooperation and collaboration are far more effective than competition when the environment is changing and resources are scarce. He was thinking of Russian wheat fields and bad weather, but his insight applies equally well to the world today.

The Earth is sensitive. Changes in atmospheric gases affect all of our planetary systems. Without carbon in the atmosphere, we are a frozen Mars, and if too much, the cauldron of Venus. We are one of 8.7 million species on an exquisite, delicate planet. In sheer biomass, human beings represent 0.01 percent of all life. All other life-forms create bountiful entangled communities that do not double-glaze the atmosphere with carbon dioxide. To better understand the community of life, we need to look no further than the community of our bodies. Our bodies would perish without untold trillions of microorganisms that live within and upon us. Each cell conducts millions of activities at any given moment. These occur because flows of carbon connect, integrate, and interact flawlessly. This is our planet, and this is our body, intricately fused to its complex home. Your body's collective cells undergo ten times more processes in one second than there are stars in the universe. This was foretold by Charles Darwin when he poetically predicted that science would discover that each living creature was a "little universe, formed of a host of self-propagating organisms, inconceivably minute and as numerous as the stars of heaven." Life only exists in cells, each cellular community containing

100 trillion atoms that self-organize into molecules that create and maintain the conditions essential for existence. When cells clump together, which they like to do, they form the biological galaxy of the human body and the species we share the planet with, from corncrakes to protozoa, smelt to crickets, blue whales to calendula.

The vanity of the solitary, self-sufficient individual exists solely in comic books and Westerns. Most aspects of modernity amplify that delusion. From legal rights to deeds, from economic theory to the right to own an assault rifle. We are urged to fight and combat climate change, a delusional reprise of Don Quixote, a stark example of how we "other" the living world. To change the atmosphere, we will need to mimic the intricate flows of planetary carbon. Social and economic relationships need to be integrated within rejuvenated social and natural ecosystems in ways that concentrated forms of economic power cannot overrule.

Western science became the dominant basis for classifying the living world in the Age of Enlightenment. Plants were things, forests were cellulose, fungi were food, soil was dirt, animals had no feelings, and nature was there to be extracted, commodified, and sold. It was a profound failure of imagination and perception. The curiosity and ingenuity that sparked the Age of Enlightenment became scientism, an unswerving rationalism that dismissed other ways of knowing. It observed nature and developed testable models that putatively explained the natural world. Except they didn't.

Original inhabitants who lived continuously on the same

land, sometimes for over fifty thousand years, see nature differently. The living world is a family, and as with all relations, a life that never repeats itself. The presence and survival of some five thousand Indigenous cultures depended upon their becoming masters of pattern recognition to understand how to thrive in forests, deserts, the Arctic, islands, and grasslands. Their teachers included all that thrived: plants, animals, elders, children, and those that came before. Native Americans gathered, hunted, and farmed in ways that created bountiful food and resources for those who followed. That Westerners do not act or see themselves this way reflects what we believe, what we have done, and what we are coming to regret.

The human journey is the daily practice of gaining and sustaining life. We can do this selfishly or gracefully. Within and around us is a living, breathing sphere of consciousness woven by a billion years of evolving life. Sentience is underfoot, in the canopy, in the favelas, in the breath of a child, the intricate, masterful web of life beneath, above, and around us. This awareness is always our story. A broken planet lies before us, but there is also a buzzing, thrumming, thriving sphere imbued with imagination, mystery, and courage. These pages are a journey into the realm of plants, the cosmos of insects, labyrinths of fungi, droves of mammals, spinneys of trees, and convocations of human brilliance. The flow of carbon is a sacred dance that entwines and weaves through all our stories.

Philosopher and Yoruba poet Báyò Akómoláfé describes moving away from the profane and desperate toward a deep sense of respect for our sacred home: "May this decade bring more than just solutions, more than just a future—may it bring words we don't know yet and temporalities we have not yet inhabited. And may we be visited so thoroughly and met in wild places so overwhelmingly that we are left undone. Ready for composting. Ready for the impossible."

TWO

Elements

Not only is the Universe stranger than we think,
it is stranger than we can think.

WERNER HEISENBERG

Carbon is the most mysterious element of all. It forms molecular chains that capture energy and retain memory.
Only one element in the universe can do that. It provides the
structural framework for trees, cells, shells, hormones, organ-
elles, eyelashes, bones, and bat wings. It is the engineer and
maker, a molecular agent animating every trace of life. Carbon
organizes, assembles, and builds everything everywhere, from
reefs to rhinos, plants to planets. The hide, scales, and membranes that enfold and protect life are made of carbon. It en-
ables and informs every aspect of consciousness, a benign
sovereign directing the expression of the living world. It can do
so because it connects and disconnects, holds fast (coal), releases easily (sugar), is pliant (bamboo), and shimmers in the

cornea of a cheetah's eye. Carbon is the keystone element of sentience, the caretaker of DNA, and the bard of the mitochondrial battery that releases the sun's energy, aka starlight, into our bloodstream. Organisms share and swap carbon promiscuously, assembling near infinite life-forms, one of which is *genus homo*, the primate that learned to walk on two legs and master fire. Carbon's incalculable manifestations make it the currency of abundance, the central bank of evolutionary growth, and the most socially adept entrepreneur in the pantheon of life. It combines nitrogen, oxygen, and hydrogen to form amino acids, the starter kit required to assemble a protein. The food for every living being, whether bacteria or elephants, is a carbon compound—fats, fiber, proteins, carbohydrates. When we digest, we break down carbon molecules, rearranging them into blood, genes, hormones, and fuel. Food begins when light meets leaf, transmuting carbon and oxygen into sugars and cellulose.

Those who call carbon a pollutant might want to lay down their word processor. No matter what we believe or betray, carbon-based molecules have the last word. This is a good thing. We are an uncommonly new species on the planet with a rather unusual brain, prone to extraordinary errors of judgment, still getting our feet wet when measured against geological time. Nature, on the other hand, never makes a mistake. It experiments, one of which is us. As long as the sun shines, the flow of carbon on planet Earth moves life toward increasing complexity, abundance, and beauty. Homo sapiens are the only species that blocks, subverts, and breaks the flow of carbon.

The prediction that increased atmospheric carbon dioxide levels would warm Earth's atmosphere was discovered by a pageant of scientists in the nineteenth century. French chemist Joseph Fourier calculated that the Earth was warmer than it should be given the heat hitting the planet. In 1824, he theorized that gases in the atmosphere must trap heat. In 1837, he predicted that overall levels of warming could change depending on human behavior and activity. Eunice Newton Foote followed, in 1856, an amateur scientist and women's rights activist who did the first experiments with sealed carbonic acid in glass bottles placed in the sun. Her conclusion: the results measured in her glass jars could happen to the planet: "An atmosphere of that gas would give our earth a high temperature." Her work was ignored for nearly a century because scientific gatherings did not allow women to speak. Three years later, Irish scientist John Tyndall devised precision experiments that showed that water vapor and carbon dioxide trapped heat remarkably effectively, nearly one thousand times greater than plain dry air. Tyndall was "baffled" by the results of his experiments and repeated them over and again. In 1896, Swedish physicist Svante Arrhenius got down to business and, using new data, spent a year calculating by hand what would happen if the amount of atmospheric carbon dioxide doubled. His answer: the Earth would be five to six degrees warmer, a number that has held up to this day using the most powerful supercomputer models. Arrhenius believed global warming would proceed slowly over centuries and could be a positive transformation for the world, especially for people living in colder climes. He could not have imagined the exponential growth of carbon emissions in the twentieth century.

Unto itself, carbon is simple. The word's root is the Indo-European term *kerh*, "to burn," which later evolved into the Latin word for coal, *carbonem*. Carbon has atomic symmetry: six protons, neutrons, and electrons. Four of the six electrons are available for sharing, either carbon-to-carbon or to other atoms, making the element available, loyal, and fickle in its versatility. A bond can be inordinately fixed, as with the lattice structure of diamonds where carbon binds to carbon; or easily broken, as in sugars where carbon bonds to oxygen and hydrogen; or somewhere between, as with sticks and skin. Although it makes up an astonishingly small percentage of the universe, barely one in one hundred thousandths, carbon is found in 90 percent of the molecules in interstellar clouds and 99 percent of the thirty-three million substances on Earth. If carbon were an animate being, we would praise its social intelligence, its gregarious, congenial, and flexible nature, and how easily it makes friends. If a carbon molecule is minutely analyzed, it is a particle zoo: leptons, quarks, bosons, mesons, baryons, and other subatomic particles that last infinitesimal fractions of a second appear and disappear. Yet, we can scrape carbon off the bottom of a cast-iron frying pan or smear a burnt candlewick on our forefinger.

Carbon has traveled to and from the atmosphere for billions of years, but not at the rate seen in the past century. Humankind created a new geological era by burning ten million years of fossilized carbon in a few centuries. Youth, the more recent arrivals to the planet, are astounded that preceding generations understood the peril of rising levels of greenhouse gases and did nothing. It was not discussed at home, work, school, or in the media. The cone of silence is receding as extreme weather vexes more people. Yet most of us still say little, go to work, clean the yard, till our fields, drive carpool, go to the factory, buy groceries, and watch our screens. It is not a water-cooler subject, even as the viability of civilization is in question. Is humanity unfazed or uninterested? The media favors celebrities, scandals, peccadilloes, and sports rather than the demise of oceans, forests, lands, and peoples. That is not a surprise. I glance at those useless stories sometimes. I want to be distracted, too. A steady diet of devastating news craters mental well-being. Our minds are not equipped to deal with the descriptions of what may befall us or others. The extraordinary economic system that had worked so well is destroying its creators. How could that be? It seems implausible. There is no vantage point where one can step back and draw a bead on carbon or climate. Both are invisible. We see weather patterns, not climate. We see people but not the mitochondria that power them. We hear frightening news but rarely a credible way forward or a meaningful path for personal action.

The result is that over 99 percent of humanity does little or nothing about the climate crisis. We are numbed by the science, puzzled by jargon, paralyzed by predictions, confused about what actions to take, stressed as we scramble to care for our family, or simply impoverished, overworked, and tired. Whatever the causes—incapacity, ignorance, or apathy—the prospect of an impending hell does not sell well or inspire. As I write this, climate activists from Extinction Rebellion in the UK have gone to elementary schools and told young children that they are doomed and it is too late. In 2021, an international survey of youth ages sixteen to twenty-five revealed that 56 percent believe humanity has no chance. People need a path of understanding that does not knock their breath away. Most of humanity doesn't talk about climate because we do not know what to say.

The cavernous, psychological dark hole called the climate emergency is occupied primarily by scientists, activists, the media, a sprinkling of politicians, and traumatized youth. Most of the world either doesn't know about the hole or peers down over the precipitous edge and quickly walks away. That is the state of the world regarding awareness of global heating and its impacts. The climate message is toothless because it is missing the core truth. Science writer Matthew Shribman believes we are witnessing the biggest communication failure in history in that the climate emergency is all about the natural world. Shribman describes the largest animal migration on Earth, an extraordinary flow that occurs nightly in the ocean. Among the billions of fish, shrimp, and mollusks are copepods, tiny crustaceans about one sixteenth of an inch long that live in bogs, swamps, lakes, puddles, caves, ocean trenches, leaf litter, and streambeds—virtually anyplace water is found, including little pools collected by leaves and blossoms. In the ocean, their nightly migration to the food-rich surface displaces water equal to the tidal movement from the moon. On the nighttime surface, hidden from predators, copepods nibble on minuscule phytoplankton that capture atmospheric carbon. As they descend again at dawn, their carbon-rich poop falls to the ocean floor, deposited for tens of thousands, if not millions, of years. The oceanic surface layers where they live are estimated to be the most significant carbon sink, two billion tons of carbon per year, just under one-sixth of human emissions.

How do we enhance and increase the flow and stock of carbon? Maybe begin by removing an inversion layer in our thinking. The overwhelming array of problems we face smothers what are possibilities, even the possibility of possibilities. Every problem is a solution in disguise, or it would not be a problem in the first place. We can start by realizing that we experience carbon's physics, biology, and biochemistry every moment we touch, taste, breathe, see, walk, converse, and imagine. There are approximately 1.2 trillion carbon atoms in every one of our 28–36 trillion human cells, busily engaged with your life. In addition, science estimates that human beings contain another 40 trillion microbial cells that coat, embed, and suffuse every body part, from incisors to intestines. The human body cannot live without them. The microbial world gave birth to us billions of years ago and never left. What do microbes have to do with the climate crisis? They keep the world alive. Science estimates that less than 1 percent of the microbial population has been identified.

The climate crisis and the intricate flows of carbon are inextricably linked. The amount of carbon added to the atmosphere since the industrial age is a pittance compared to the total carbon stock on earth, less than 0.00004 percent. It illustrates how sensitive the atmosphere is to additional carbon. Doubters disbelieve that such changes could make a difference, which makes sense. How could going from 280 parts to 425 parts per million of carbon dioxide in the atmosphere make much of an impact? It seems minute. Humans are no less sensitive. The hormones estrogen, testosterone, progesterone, cortisol, insulin, and melatonin, which determine our mood, weight, libido, sexuality, sleep, and metabolic health, amount to no more than one hundredth of a drop of water in our body.

Human cells, and those of every animal, plant, and living organism, descended from a singular event. Two similar types of single-celled carbon-based microbes existed in the ocean for 2.5 billion years: bacteria and archaea (ar-kay-yah). They were tiny blobs with no obvious purpose. Although multitudinous, they had no scaffolding, nucleus, or internal means to generate energy. Archaea stuck to ocean vents and hot springs for sustenance, while bacteria nibbled ocean-borne amino acids and carbon compounds for their daily fare. The two microbes merged at some point, creating eukaryotes, cellular organisms that could transform light and carbon dioxide into oxygen and energy. The newly minted cells reproduced endlessly. They became intertwined and social, eventually bunching and clumping into multicellular organisms. Early prototypes were microscopic worms that evolved into eels, lobsters, trees, sloths, moths, eagles, you, and me. How did that first union happen? Despite countless attempts, science has never replicated the merger of an archaeon and a bacterium in the lab. We can trace all of life to this single mutant cell, an Edenic event Ed Yong calls "breathtakingly impossible." As organisms became complex, multi-limbed, finned, webbed, and winged, symbiosis returned the favor to the microbial world, and all organisms became habitats for microbes. Trillions are in our guts alone, and our mental and physical well-being depends entirely on their unique functions. Human bodies are communities of life without which we perish. So, too, are all forms of life. The word *community* is essential to solving the crises we face.

Can the stock and flow of carbon be tipped in our favor? Yes, if we attend to the entirety of flows—microbes to cells, fungi to plants, farms to kitchens, forests to fields, homes to communities, factories to commerce, and governance to culture. To quote Sebastian Junger, "The idea that we can enjoy the benefits of society while owing nothing in return is literally infantile. Only children owe nothing." We need to address what is down here, not up there: the massive rates of resource extraction, wealth concentration, financial hegemony, political corruption, commodification of food, cultural deracination, human exploitation, and the absurd "tragic science" of economics, which excludes the environment. We owe it to the children to speak to all causes of the crisis, areas that most climate scientists avoid or are hesitant to discuss.

THREE

Firmament

The total number of minds in the universe is one.

ERWIN SCHRÖDINGER

The sophomoric statement that we are made of stardust is technically true: the human body contains approximately forty-four octillion molecules, which is incomprehensible and objectively meaningless. If we say we are made of nuclear waste, that would be equally true. Either way, the molecules originated in the spheres of giant red stars that expand as they reach the end of their life. Our sun will do the same at some point. Its stellar core is mainly comprised of hydrogen and helium and is powered by nuclear fusion. Two hydrogen atoms, the most abundant element in the universe, fuse to become one helium atom (known as an alpha particle), the second-most abundant element in the universe. It is not a perfect match. An unneeded neutron spins off when the atoms fuse, generating vast amounts of energy (nine ounces of hydrogen could power the global economy for a month). When sunbathing at the beach, it would be accurate to say you are star-bathing.

Over billions of years, the intense heat and gravitational pressure at the core of a star create a torrent of nuclear reactions that engender successively heavier elements: oxygen, nitrogen, sulfur, sodium, and more. The creation of each element releases ever more energy, stoking a solar bonfire and increasing the rate of elemental transformation. That is until the twenty-sixth element emerges—iron. The protons, electrons, and neutrons that comprise iron are reactive, but unlike the lighter elements, the reactions absorb energy and release none. When a star reaches a critical mass of iron, energy generation ceases. The fusion reactor shuts down, the system flips, the heating is gone, and within a quarter of a second, a once brilliant star collapses into a majestic supernova that is visible for months. Every second, somewhere in the universe, stellar eruptions blast elements into space, forming nebulae, cosmic dust clouds trillions of miles wide containing sulfur, argon, cobalt, lead, graphene, and gold. To grasp the immensity of ongoing creation in the universe, imagine eighty-six thousand suns exploding daily. These immense dust clouds spanning hundreds of light-years shield their atomic nursery from ultraviolent destruction and journey through the cosmos for hundreds of millions of years, creating molecular permutations from the castaway elements. Gravity eventually forces interstellar clouds into elaborate vortices of gases, dust, and pebbles. As the compressive spiraling forces of gravity increase, a flattened disc of a new sun is formed, encircled by a chaotic mixture of rubble that eventually coalesces into planets. We are the progeny of a death star, "a sheaf of empty space and ancient electricity, an unimaginable quantity of atoms, carrying protons and neutrons and somersaulting electrons." Stars beget stars. And us.

In the 1950s, the prevailing creation theory assumed that the atomic elements were created during the Big Bang. Astrophysicist Fred Hoyle coined the term *Big Bang* in a BBC interview as a term of derision. For Hoyle, the idea that a universe detonated itself into being and cooked up the entire suite of elements in a brief moment was absurd. As far as he was concerned, the universe had no beginning. He believed in a steady-state model where an expanding universe constantly created new stars and galaxies, which was in accord with his atheism. The steady-state theory of the universe was eventually tossed out. The creation of the universe was traced back to a single point in time, a singularity occurring approximately 13.8 billion years ago. In 2023, images from the James Webb Space Telescope, peering back 13.5 billion years, revealed fully formed galaxies that may not have been created within that timeline.

From the point of view of Hindu or Buddhist cosmologies, the Big Bang and steady-state theories are both correct. Compared to Western cosmologies, ancient teachings propose unimaginable time scales wherein the universe expands and contracts repeatedly during millions of maha kalpas. The mythical duration of a maha kalpa is how long it would take for a mountain three times higher than Everest to be worn down to dust by a dove flying above and rubbing a silk cloth over the peak every

100 years, approximately 311 trillion years. It is an allegory of infinity. In this cosmology, the Big Bang was simply the most recent of infinite pulsations of the universe.

Regardless of cosmological theories, scientists were puzzled about the existence of carbon. If carbon was not created in the Big Bang, it had to be made in stars, but how? It was first assumed that three alpha particles (helium) could be fused into a carbon atom. An alpha particle has two protons and two neutrons. Multiply times three, and you get carbon with six protons and six neutrons. The math was perfect, but there was not the required "energetic state" to achieve fusion. Nuclei hold a certain number of protons and neutrons. Their configuration around the nucleus determines the amount of energy contained. Different energy states, like staircase levels, arise when protons and neutrons are arranged differently. The variety of energetic states in atoms are referred to as "resonances" by physicists. Three alpha particles cannot become carbon due to incompatible resonance.

Atoms are virtually empty spaces that repel other atoms with extraordinary magnetic force. If they did not repulse, fusion would occur when we stepped on dirt, and the Earth would be no bigger than a baseball. The question remained: How did lighter elements become carbon? Hoyle was intrigued by the puzzle. "Since we are surrounded by carbon in the natural world, and we are ourselves carbon-based life, the stars must have discovered a highly effective way of making it, and I am going to look for it." You might say he did the math. He knew that two alpha particles fuse into beryllium, an extremely un-stable atom that quickly reverts back to alpha particles. In physics, quickly has a different meaning. The span of time in which two helium atoms fuse and could be joined by a third alpha particle was calculated to be one trillionth of one tril-lionth of a second (0.0000000000000000968). Hoyle pro-posed that if a third alpha particle collided with the unstable beryllium in that infinitesimally small moment, the reaction would alter the resonance of the elements and fuse into carbon.

Looking back at that time of debate and uncertainty, astro-physicist and journalist Marcus Chown called Hoyle's idea "the most outrageous prediction" ever made in science, another way of saying his peers considered it absurd. In 1953, Hoyle gave a series of lectures at Caltech on stellar physics and how elements had synthesized from helium to iron. Scientists at Caltech treated his thinking with deep skepticism. He was an astrono-mer speculating about nuclear physics. Ward Whaling was ini-tially dismissive of his lectures, saying it sounded like he "was making things up as he went along." Others rebuked his theo-ries of stellar properties outright. Although some thought his theory laughable, Hoyle remained adamant, asserting that there was no other conceivable pathway where carbon could have been created. He pressured scientists to test his theory.

Skepticism of Hoyle's theory was amplified by the fact that no spectral analysis of gases in the universe had ever detected his imagined form of energized carbon. Having been badgered relentlessly by Hoyle, nuclear physicist Willy Fowler, one of

Hoyle's skeptics, eventually formed an experimental team to see if they could create Hoyle's energized carbon. Fowler and a team of some of the world's brightest nuclear physicists lined up to test the theory. They had to move a several-ton mass spectrometer containing its giant magnet down a four-foot corridor with two right angles. A team of graduate students placed the steel plate upon hundreds of tennis balls, slowly advancing the device by placing the tennis balls continuously in front. Led by Ward Whaling, who believed Hoyle was fabricating data, the team worked backward, bombing nitrogen with hydrogen isotopes to remove protons and create carbon. After three months of intense work, Fowler's team validated Hoyle's outrageous prediction, an enhanced state of carbon. Whaling and his team submitted a paper to the American Physical Society, placing Hoyle's name first. The estimation and regard for Hoyle changed drastically at Caltech, and Fowler and Hoyle became friends for life.

Once it was shown how alpha particles (helium) could be turned into carbon, it demonstrated how ascending additions of alpha particles created other vital elements. Add two protons and two neutrons to carbon-12, and you get oxygen-16, and so on to create neon-20, magnesium-24, silicon-28, argon-36, and calcium-40. Working the other way, when two protons break away from oxygen-16, you have nitrogen-14. Amino acids comprise three elements—carbon, oxygen, and nitrogen. Hoyle didn't just predict the mechanism of carbon synthesis; he identified how the essential elements of life originated and how building blocks were created in devolving massive red stars. When a star eventually collapses into an exploding supernova, carbon and its progeny of elements are strewn throughout the universe. The molecular seeds are disseminated like wind-borne dandelion florets.

The primordial state of entropy in a dying star is now known as the Hoyle state. In the Hoyle state, only one out of approximately twenty-five hundred fleeting and oddly formed beryllium nuclei survive and become carbon in that one trillionth of a trillionth of a second. And yet, carbon is prolific. The mantle of Earth contains an estimated 1.85 billion, billion tons of carbon. The atmosphere is another 585 billion metric tons, the soil is 2,500 metric tons, and plants are 450 billion metric tons. Hoyle had previously scoffed at the idea that a God or higher force existed in the universe. Life was an accident, a random concurrence of atoms, molecules, and chemistry, and to see the universe any other way "was a desperate attempt" to escape reality. His discovery changed him. "A common-sense interpretation of the facts suggests that a super-intellect has monkeyed with physics, as well as with chemistry and biology, and that there are no blind forces worth speaking about in nature. The numbers one calculates from the facts seem to me so overwhelming as to put this conclusion almost beyond question." It was not beyond question to his peers. Hoyle lost credibility in the scientific world with a statement that hinted at forces (intelligence) beyond our comprehension. That was undoubtedly why he was excluded from the 1983 Nobel Prize awarded to

Willy Fowler, who proved Hoyle's prediction even though he had initially doubted it.

Today, when physicists describe the kind of coincidences required to create the basis of life, they speak of extraordinary symmetries, resonances, gradients, and values, from the smallest subatomic particles to the sun's gravitational field. According to University of Cambridge physicist Stephen Meyer, physicists have "discovered that life in the universe depends upon a highly improbable set of forces and features and an extremely improbable balance among many. The precise strengths of the fundamental forces of physics, the arrangement of matter and energy at the beginning of the universe, and many other specific features of the cosmos appear delicately balanced to allow for the possibility of life. If any one of these properties were altered ever so slightly, complex chemistry and life would not exist."

Physicists have given a name to a universe that features a dozen or more implausible coincidences, a name that is more reminiscent of a race car engine than intelligence: "fine-tuning." One might say that the resonances and impossible harmonies imagined, deduced, and finally proven by Hoyle, Fowler, and Whaling could be compared to a symphony, which requires that voices and instruments merge at identical and complementary frequencies at precisely the exact moment. The first performance of Gustav Mahler's eighty-minute *Resurrection* Symphony required 858 singers and 171 instrumentalists—all in perfect tune and pitch, playing and singing under one conductor with extraordinary consideration and connectedness. If you have never heard it, listen to the last movement. The chances of the universe creating the resonance and alignments conducive to the orchestra of life are calculated by physicists to be virtually incalculable. And yet here we are, me writing and you reading.

FOUR

Cell Mates

*Even today, we can't really explain what the
difference is between a living lump of matter
and a dead one.*

SARA IMARI WALKER

When I was four, vernal pools of salamanders, shuffling
frogs, and scuttering tadpoles were practically family. I
would sit by a creek and watch striders and water boatmen endlessly. One day, I chased a ball onto the street and discovered a
frog flattened like a tortilla. There were tread marks embossed
on its spots and lumps. The splayed reptilian legs pointed in all
directions. I was sure the frog had been run over by the ice
cream truck that circled the neighborhood playing the "Cuckoo
Waltz." The dried-up carcass was as thin as cardboard and
light as a wafer. I peeled it off the road and ran to show my
mom. She threw it in the trash and washed my hands. I went to
the garbage can later and placed the tortilla-shaped frog into

my treasure box under the bed, which included vacant snail shells, tattered butterfly wings, and detached tails from lizards that had escaped my grasp. Despite my creature loot, I hadn't fully realized that something could die. It was a revelation, and I did not like it one bit. I became obsessed and anxious. It meant my sister, dog, and mom could die. On a moonlit night, I was awakened by the sound of a frog spilling down from above. I ran outside in pajamas, and there on the roof was a bird whistling, chirping, singing, and trilling. The bird imitated crickets, crows, jays, my grandfather's squeaky weathervane, and frogs, bobbing its head up and down as if to say, "I am having the best time in the world up here." I thought I was in the presence of God. Seriously. When you are young, the divine seems nearby. Who knew she was a bird, for who else but God would sing and dance joyously all night on a corrugated tin roof? In the morning, I ran excitedly to find my grandpa to share the news. He listened politely, nodded, paused, and said, "Son, that was a mockingbird."

In school, we were taught that life is a competitive struggle. The word *cooperation* was not mentioned in science class and was seldom seen in the school. We were graded on a curve, not as a team. We were taught Darwin, not Saint Francis. Playground insults, cattiness, hazing, and the occasional pummeling verified the classroom tutorial. In the background was the nightly news detailing convulsions of regional wars and economic tumult. Yet, I never saw conflict in the apple trees, the creek by the house, or among the crows that gossiped at dusk in the pine trees. There seemed to be two worlds.

Biologists once clearly defined life. It has movement, reproduction, metabolism, energy, perception, membranes, and organization. These attributes were said to be common to all living organisms regardless of shape, species, or size. Due to recent biological discoveries, that clarity is no longer possible. Because there is no longer an accepted definition of life, biology remains the only science that cannot state precisely what it studies. This is in contrast to one's experience. Our perception of life is instant and instinctual, as it is for all creatures. Cells contain trillions of molecules that interact chemically. Whether we are talking about a microbe or a manatee, the cells of all organisms are hotbeds of carbon. The molecules in cells are lifeless, metabolic tool kits for their intricate world. How do trillions of inanimate molecules in a single cell become sentient? None of the molecules are alive, yet the cell is a living organism, a phenomenon yet to be explained.

Over the past decades, biological research has expanded the scope of the living world. In 1969, extremophiles were discovered, microorganisms found in seemingly unlivable conditions that survive levels of heat, cold, and pressure that no previously known life-form could withstand. Extremophiles reside in lakes half a mile under the Antarctic ice sheet and are found in rocks buried nineteen hundred feet beneath ocean seabeds.

Tardigrades are micro-animals known colloquially as water bears or moss piglets. They look like obese caterpillars with a squashed visage of an alien. Tardigrades have existed for over five hundred million years and can withstand nuclear radiation, starvation, oxygen deprivation, and bone-crushing pressure.

When completely dehydrated, they look like sesame seeds, with no legs. Decades later, if you sprinkle them with water, stumpy little legs reemerge, and they waddle away.

Rotifers were first called animalcules, a term coined by Antonie Philips van Leeuwenhoek, a master lens maker from Delft in the Netherlands. Known as the father of microbiology, Leeuwenhoek's initial scientific contribution was precision microscopes with a high-powered lens. They were ten times more effective than the best microscopes of his day. His second contribution to science was to look where no one had thought to go. In 1674, he placed a drop of rainwater onto a glass plate and became the first person to observe microorganisms: bacteria, protozoa, archaea, fungi, and algae darting about. "No more pleasant a sight has ever come before my eyes," wrote van Leeuwenhoek. It was pleasant to him, but he was afraid to share what he saw with anyone else. At the time, his discovery was not credible. The microscope master with a grade-school education eventually got up the nerve to send a letter to the Royal Society in London. Overcoming deep skepticism, a delegation arrived and saw living, pulsing microbes with their own eyes. It transformed the world of science. Years later, when examining a droplet of water from his lead gutters, van Leeuwenhoek found rotifers, twin-headed blobs of zooplankton adorned with a small paddle tail. In the heat of the summer, he mixed gutter dust with water and saw rotifers come back to life. In 2021, a scientist retrieved a rotifer from the Russian Arctic between 23,960 and 24,485 years old. When thawed, the ancient rotifer started to reproduce.

Viruses run circles around tardigrades and rotifers but are not classified as a life-form. The International Committee on Taxonomy of Viruses states that viruses are not living organisms. Since they do not live, we should use the word residents; they reside in living cells and transform them, sometimes for the better, sometimes for the worse, but mostly in ways we do not fully understand. Viruses are tiny compared to a cell, like a small child standing in front of the Empire State Building. They move in and out of organisms and plants with ease. They neither eat nor grow. Viral genetics change the structure of cells, producing millions of duplicates and occasional mutations. This is how COVID-19 spread and evolved. It is estimated that there are over one trillion species of viruses. Billions suffuse our gut, skin, eyes, hair, mouth, and organs. Viruses have a reputation. Although some viruses such as smallpox, polio, and Ebola can seriously harm human health, scientists have a different narrative. Without them, life would not exist. Inside our body, their collective presence is called the virome. Carl Zimmer writes, "If viruses are lifeless, then lifelessness is stitched into our being."

In 1992, NASA began preparations to search for life in the universe, within and beyond our solar system. It created a department of exobiology where microbiologists evaluated which detectable chemicals would signify the presence of life. There were multiple definitions in the biological community, but NASA required a clear working definition to ascertain extraterrestrial life's presence. No institution studied the problem more intently than NASA, and it did so with an all-star cast of biologists. The scientists were in broad agreement: Life extracted energy from the environment, made copies of itself, had membranes, responded and reacted to the external world, metabolized matter and shed waste, needed water, and was carbon-based. NASA needed to anticipate life-forms different from anything known on Earth. After much discussion, biologists came up with an all-encompassing definition of life as a "self-sustaining chemical system capable of Darwinian evolution." With that bit of Earthly bias, NASA set out to find it.

In 2011, Edward Trifonov, a Russian geneticist of some renown, hoped to unify the divergent life perspectives. He assembled 123 definitions of life, including NASA's, hoping there might be a primary underlying theme that would unite them all. As an expert in protein structure, he believed a common thread lay within the linguistic structure of the multiple definitions. He took the seven-word NASA definition and boiled it down to three: Life is "self-reproduction with variations." His description was a triumph of brevity. It seemed to encompass all 123 extant descriptions. However, it was a short-lived victory. Under his definition, computer viruses were a life-form.

Religion has a similar problem. There are hundreds of names for God: Yahweh, Jehovah, Elohim, Olodumare, Creator, Almighty, Hu, Bahá, Allah, Bhagavan and Bhagavati, Nana Buluku, Chukwu, or Unkulunkulu. As with the varied definitions of life, the descriptions of God do not differ widely. However, they do not create a shared understanding. Whereas religion extends to the invisible world, life stays home and cavorts with matter and molecules. In physics and chemistry, there are principles and constants that are inviolable. In biology, the agreed-upon principles that determine life are under discussion.

Physicists have identified successively smaller subatomic particles that make up the elemental world. Biologists do the same to the cellular world. However, in neither case does minutia explain an atom or a cell. Is carbon interlinked to the array of living processes in a way we cannot imagine? The quick answer from physics is an unqualified no. Biologists would agree. The flow of carbon does not belong in the conversation other than as an all-purpose element. Yet physics, like biology, cannot draw a bright line between the animate and inanimate. Physical sciences divide, isolate, and separate, trying to understand the whole of life through its smallest parts. Life is none of the above. Einstein once remarked that our perception of being separate from life is an optical illusion of consciousness.

The best way to understand life may have been proposed by a philosopher of science, Carol Cleland. She is convinced that defining life is the wrong way to go. Definitions are for dictionaries, not the living world. Cleland believes we need to understand the nature of life, not life in nature, and that entails understanding life as a system, whether ecosystems, colonies of extremophiles in sulfurous ocean vents, or microbial communities in our gut. I would add Indigenous cultures to that list, for they, too, are enduring living systems. And why not Athens and Kyoto and other venerable cities, which are also long-standing living systems?

In 1961, famed English scientist James Lovelock worked as

a consultant to NASA to develop highly sensitive instrumentation that could detect the presence or absence of life in extraterrestrial atmospheres and surfaces. This began a planetary exploration program that would send probes into space. The first goal was to send a two-part spacecraft to Mars that would photograph the landscape from above before landing and analyzing dirt samples below.

Besides being the second-closest planet, Mars had been the object of endless speculation after Italian astronomer Giovanni Schiaparelli, director of the Brera Observatory in Milan, announced in 1878 that he discovered *canali* on the planet, the Italian word for channels. The term was mistranslated into "canals," implying the presence of civilizations. In 1895, Percival Lowell observed the same straight-line tracks and believed them to be how water from the poles was directed to equatorial regions. His imaginative interpretations of what he saw described "unnatural features" and oases where canals crossed, which furthered his speculative writings that Mars was an "abode of life." Subsequent extraterrestrial life-proposing literature were classics like *The Martian Chronicles* by Ray Bradbury, *Stranger in a Strange Land* by Robert Heinlein, and the famous 1938 Halloween adaptation of *The War of the Worlds* by H. G. Wells on CBS radio that created panic for listeners who thought it was a live broadcast of the Martian invasion of New Jersey. People had long-forsaken fantasies of hidden worlds within the red planet when the *Viking* mission was announced. Still, curiosity about whether there was a vestige of life remains high.

NASA's *Viking* program sent two space probes to Mars in 1976 on staggered launch dates. Lovelock's instruments and detectors were on both. After photographing and scouting for ideal sites, both orbiters released landers containing little shovels, miniature backhoes that scooped up dirt placed into ingeniously designed laboratories. The labs had three different "soups" that would provide nutrients to hoped-for fungi, microbes, or extremophiles. After a few hours, the off-gases were analyzed, and the results were relayed back to scientists at NASA.

The test results offered no conclusion. A planet with life would have an atmosphere that would be a chemically dynamic mixture of reactants and gases. The tests showed the Martian atmosphere to be stable and virtually dead. It is almost entirely carbon dioxide with minute amounts of oxygen, methane, and hydrogen, which means no life-forms are present. The question remains: Did life ever exist on Mars?

In September 2022, a month after James Lovelock passed away at age 103, NASA announced that the *Perseverance* mission to Mars had collected four sedimentary rocks from an ancient river delta that might prove that life once thrived on the planet. The US and European space agencies plan to send spacecraft to the Jezero Crater to pick up the rocks and hope to return them to Earth for analysis by 2033.

Before the *Viking* launch, Lovelock knew that life-forms interact with the atmosphere and change its composition. Lovelock and some colleagues had been researching the complex atmospheric influences of phytoplankton blooms in the ocean.

Rather than seeing the atmosphere as the result of biological exhalations, they concluded that phytoplankton was altering and adjusting the atmosphere through negative feedback loops. As the atmosphere warmed, phytoplankton growth increased and captured more carbon dioxide. When there was cooling, the phytoplankton population scaled back. This was symbiosis, a mutually beneficial relationship between two organisms—except the atmosphere is not an organism. However, their research made sense if the complex interactions and feedback between the biosphere and atmosphere were seen as a single organism. Eric Roston writes, "The Earth is essentially a closed material system. The amount of carbon, water, and other materials is probably about the same as when the planet formed. From this perspective, evolution is an expanding-contracting regulator of the path of carbon through the Earth system, rewiring . . . the atmosphere, the oceans, and the land."

From his and adjoining research from Alfred Whitehead and Evelyn Hutchinson, Lovelock developed the Gaia hypothesis. It proposed that the Earth behaved as a single living entity instead of a planet of divergent and competing entities. Nobel Prize–winning author William Golding suggested the name Gaia, after the Grecian goddess who was the mother of Earth and all life.

Although there are variations of the Gaia hypothesis, the core version posits that coordinated interaction between living organisms protects the planet for the benefit of the whole of life. Gaia was not a popular notion when the hypothesis was first presented. One scientist scornfully called the theory "New

Age pablum." Other criticisms centered around the behavior of organisms. Because species act in their self-interest, Darwinian natural selection and selfishness would preclude collective impact. The counterargument notes that creatures, including humans, are alive because of homeostasis, a dynamic equilibrium where self-regulating processes maintain internal stability. If your body deviates from homeostasis, you are in trouble, and it can lead to disease and death. Our self-regulating processes include chemical and physiological optimizations that regulate body temperature, fluids, and blood sugar. Lovelock saw the planet as having the same characteristics, the tendency toward stability among seemingly independent elements, including the atmosphere. His observation was simple: the Earth can "regulate its temperature and chemistry at a comfortable, steady state." How could the immeasurable population of life-forms on Earth collectively create planetary homeostasis? Fair question, which asks another question: Does the atmosphere maintain homeostasis for the biosphere? Or the opposite? Gaia's theory says yes to both and provides a framework of understanding that does not align with Darwinian paradigms.

The inference from the Gaia hypothesis is straightforward. Do not harm the mantle of the Earth or the seas. Restore the planet's biological metabolism that has been defiled by years of mining, deforestation, poisoning, overfishing, acidification, and habitat destruction. Animals are slaughtered for their tusks, hides, horns, and meat. Much of the animal kingdom is debilitated by buildings, highways, farms, lights, noise, chemicals, and habitat loss. We are saying goodbye to vaquitas, river dolphins,

giant pandas, Amur leopards, curlews, mountain gorillas, forest elephants, maned wolves, and even the once common house
sparrow.

There are influences we are less aware of. To protect the
planet's life, we must put away our weapons, long lines, chainsaws, toxins, bulldozers, drilling rigs, and starter mansions. We
need to be quiet and turn off the lights. Intrusions of bright
light and confusing sounds transform the world of birds, insects,
and mammals. Sound and light bewilder and overwhelm the
senses of species that hear, touch, see, and know the world differently than we do. LED headlights hurt even our eyes, but
more importantly, they distort the night and wipe out the
moon. Firefly populations plummet in the face of light pollution. Imagine how a mall-sized parking lot would appear to a
bat, owl, or nightjar: a feast of insects circling high-intensity
LED lights that erase the night. Two thirds of invertebrates
are active at night. Insects hunt, pollinate, and mate using the
moon and the stars as navigational guides. Moths sense the
moon and Milky Way upon their back, which allows them to
fly directly to their destination. When insects encounter artificial street- or porch lights, they become hypnotized and fly in
circles. The insects not picked off by bats often flutter into exhaustion and death. Owls will swoop in and grab the bats.

The terrorist assault on the World Trade Center is memori-
alized on September 11 in New York City by two columns of
intense light symbolizing the Twin Towers. Three hundred
thousand watts comprised of eighty-eight xenon searchlights
extend several miles into the atmosphere visible sixty miles

away. Known as the *Tribute in Light*, the weeklong display attracts and confounds over a million migrating birds, including redstarts, orioles, vireos, pine warblers, swifts, and flickers. Ed Yong writes, "Migrations are grueling affairs that push small birds to their physiological limit. Even a nightlong detour could prematurely sap their energy reserves to a fatal effect." Scores of birders take shifts during the night to monitor bird behavior. If a bird crashes into a building, if they seem lost or circle, or if there are more than a thousand birds around the beams, the lights are turned off so the flocks can regain their heading. Within minutes, their behavior changes, and they fly on. Other cities, such as Houston, Atlanta, and Boston, work with the Cornell Lab of Ornithology to prevent bird death. Cities that participate will black out high-rise buildings when informed about migrations. Over six hundred million birds a year perish due to building collisions.

The sensitivity shown by the *Tribute in Light* does not exist in cities, airports, high-rises, bridges, streetlamps, and porch lights. Light is a type of pollution taken in by the nervous system rather than the lungs. Yong points out, "Light has come to symbolize safety, progress, knowledge, hope, and good. Darkness epitomizes danger, stagnation, ignorance, despair, and evil. From campfires to computer screens, we crave more light, not less. It is jarring for us to think of light as a pollutant, but it becomes one when it creeps into times and places where it doesn't belong." Turtle hatchlings have evolved over millions of years on dark beaches. At night, they instinctively crawl to the horizon of the ocean, which is brighter than the land behind.

When outdoor lights come from the other direction, hatchlings pivot, move away from the sea, and may wander straight into a beach fire.

Noise is audible pollution, another type of nervous system disturbance, what Karen Bakker calls "acoustic smog" that is present at epidemic levels. It significantly impacts the soundscapes animals live within, both on land and especially in the sea, where sound travels better and farther. As with the circadian rhythm of sun- and moonlight, sound is a signal and message. Sensory information guides and directs the living world. Humans are equally sensitive to sound. What happens to us when we hear a siren, a flute, a scream, a laugh, heavy metal, or a gospel chorus? Bernie Krause, a pioneer in bioacoustics, has recorded wild soundscapes since 1979. The recordings are ecosystem symphonies of tones, notes, cries, chirps, hums, thrums, songs, and vocalizations produced by organisms within intact and undisturbed habitats. His recordings are stunning, mesmerizing, and intricate, the audible equivalent of Monet's *Water Lilies.*

To better understand what constitutes soundscapes, Krause and Stuart Gage coined terms for three categories of sound. *Biophony* is the sound of living organisms in an ecosystem. *Geophony* is the natural sound of the landscape and geology, the wind soughing through the trees or water in a stream. The third is *anthrophony*, the sounds generated by human activity. Although many of the sounds we humans create, such as music and language, are purposeful and constrained, most sound, as any city dweller knows, is chaotic, unpleasant, and uncontrolled.

Over decades of research, Krause observed a disheartening change in his recordings. Returning to alpine meadows, old-growth forests, and pristine glades previously recorded, they were different. Logging may have occurred nearby, highways or suburbs were built within hearing, or ecosystems were directly under the flight path of commercial aircraft. Krause's recording instruments showed that much of the audible spectrum in intact ecosystems is occupied, from the lower bass notes of a raccoon to the high-pitched trills of a robin at night. Anthrophony, the "human din," has spread throughout nature, polluting and suppressing the biophonic symphonies. In a conversation, Krause described partitioning, how inhabitants in an ecosystem will divide and occupy acoustical spaces in accommodation to their own and other species, just as is done for radio and broadband. Cicadas will modify and adjust crepitations, their characteristic crackling sound, to not compete with different bandwidths.

Lincoln Meadow was an intact, old-growth ecosystem 6,500 feet high in the Sierra Nevada. A logging company persuaded the adjoining community that they could selectively log a wide area instead of clear-cutting, extolling their environmental sensitivity to habitat and species. When Krause heard about this, he rushed to record Lincoln Meadow before logging began. When he returned a year later, he could still hear the geophony of the stream running through the meadow, but the sounds of biophony were almost entirely absent. He has since returned fifteen more times. Though it looks similar to what was there before logging, Lincoln Meadow never recovered.

When anthrophony converges with biophony, there is tonal chaos. When a roadway is built over a wetland, the throaty sound of downshifting diesel engines at night might occupy a similar acoustic spectrum as the low-frequency mating bellow of a bullfrog. The alpha male is now a Peterbilt truck. Once-dominant bullfrogs go quiet. Krause documented how the absence of sounds from one or more species can cause a subsequent decline in other species that share a habitat. This is similar to the metaphor of an airplane in flight, where rivets begin to pop off the wing, one by one. A few rivets are acceptable, but at a certain point, the loss of rivets will weaken the wing, and it will fail. Noise is increasing exponentially worldwide, a form of uncontrolled pollution unraveling the living world. Conversely, Krause recently returned in 2023 to sit under a bigleaf maple in a meadow he had recorded faithfully for thirty years. For the first time, there was silence. Nothing was heard. The movements in the underbrush, the orange-crowned warbler, the spotted towhee, the house wren, and the mourning dove—all absent.

Nature has listened to itself for millions of years. It acts based on those sounds, an evolving language beyond our understanding. Imagine tomorrow morning, you turn on your local radio station and you hear a garbled mixture of Iranian Persian, Wu Chinese, ranchero Spanish, and Sudanese Arabic intertwined with pidgin English, an analogy as to how creatures might hear the cacophony of sounds arising from our cities, oil tankers, roadways, and chainsaws. Aside from noise pollution, "a great silence is spreading over the natural world," according to Krause. And it is caused by a great noise. Researchers predict that by 2050, there will be sufficient roads to circle Earth six hundred times.

Although biology may not be able to define life, biologists have discovered more about the intelligence and interconnectedness of living systems in recent decades than in previous history. Starting with our body, we do not control the infinite number of events in our cells. Extraordinary fungal, viral, and bacterial networks within soil determine the health of plant life and the capacity of the Earth to store water and moderate the surface temperature on land through the hydrosphere. Trees are communities, signaling to each other, utilizing pheromones and fungal networks to advise, protect, and guide their well-being and survival. Animals have extensive communication skills and inventive minds that we are just beginning to fathom. The deep sea, one thousand feet below the surface, is the largest habitat on the planet. Over 80 percent of its inhabitants use bioluminescence to communicate, detect, and defend, making light the number one communication method organisms use on Earth. On land and sky, communication consists of bio-acoustics that impact the health and fettle of ecosystems. If you have spent a night outside in the Amazon, the most intensely diverse ecosystem on the planet, you are treated to a fauna opera. "If we matter, so does everything else," concludes Melanie Challenger in her seminal work *How to Be Animal*. "The more we take, the more their lights go out."

FIVE

Eating Starlight

If you can't pronounce it, don't eat it.

MICHAEL POLLAN

Our hunger and passion for food—its taste, aromas, color, and textures—are variations of the dance of carbon. When you detect the scent of fresh bread from the oven, you salivate at the conversion of carbohydrates into alcohol and carbon dioxide. Your bloodstream is talking to your tongue and nose. Over 99 percent of the human body is made of hydrogen, carbon, oxygen, and nitrogen. The atmosphere has the same constituents. So does food. The arrangement of those four elements determines natural and artificial flavors. Body fats, whether in your belly, skin, or scalp, are carbon, as is the olive oil in the pantry. Sugars are carbon stitched together by oxygen and hydrogen. Add nitrogen, and you have proteins that form muscle, eyes, organs, and skin. When we sip, bite, and chew, the taste and scent send neurons to the hindbrain at 100 miles per hour. That instantaneous cluster of information tells us whether to eat something. Everything you consume is comprised of carbon. We taste nutrients plants create from carbon, water, and a star.

For two million years, human beings were hunter-gatherers. They roamed nomadically from their African origins to the Middle East, Asia, Europe, and ultimately the Americas, developing tools, settlements, fire, and an intricate knowledge of plant and animal life. The diaspora is represented today by the five thousand distinct Indigenous peoples across the Earth, comprising 6 percent of world population.

The quest for food is never-ending. The abundance provided by the modern food system of domestication eliminates the need to hunt for food in most countries, but not the need to gather. Columbus was searching for spices, not an unknown continent. He landed in San Salvador seaward of the Bahamas in 1492 and then in Hispaniola, where the Spaniards engaged in "recreational slaughter"—rape, torture, beheading, and evisceration of the Taino people, the second culture he encountered. They cut off the legs of children who ran away and used nursing infants for dog food. He found some tree bark he claimed to be cinnamon to prove to Queen Isabella that he found the western route to India, where he met many "Indians."

Europeans had experienced thirty-seven famines in the preceding five hundred years, seven in Italy alone. The invaders came upon an edible landscape farmed by cultures who had not gone hungry for centuries. Unbeknownst to the conquistadors, the food discoveries were far more valuable than the looted silver and gold. Maize was developed ten thousand years ago in central Mexico. The maize Columbus brought back is cultivated today from Russia to South Africa and is the largest grain crop by weight grown in the world. Three root vegetables developed in the Americas—potatoes, sweet potatoes, and cassava—are collectively the largest source of calories in the world. When you add cacao, tomatoes, avocados, peppers, chilies, peanuts, cashews, sunflower, safflower, vanilla, pineapple, papaya, blueberries, strawberries, passion fruit, pecans, melons, cucumbers, pumpkins, butternuts, zucchini, cranberries, kidney beans, pinto beans, and lima beans, it is not difficult to concede that Indigenous farmers of the Americas were the leading plant breeders in history. We might have learned more about the history of food and agriculture in Mesoamerica had the great libraries of the Mayan people survived. In 1562, the Spanish bishop Diego de Landa burned every Mayan book (codex) he could get his hands on to suppress the pagan Mayan culture. Four mutilated codices remain, bark-paper screen folds containing history, religion, calendars, and unerring star maps.

We no longer need to seek food. Nourishment comes to us from an exceptionally complex system that has created unparalleled abundance. There are an estimated three hundred thousand edible plants, but less than two hundred are commonly used by humans. Many of the others might taste bitter, wild, or grassy, not something to put into a smoothie. However, our ancestors relied on thousands of varieties for nourishment. Today, twelve plants and five animals provide 75 percent of the human diet. The drastic reduction in diversity comes about in an era where the profound uniqueness of the human species is ever more evident. Just as each face, eye, fingerprint, and odor differ, so do our genetics, metabolism, nervous system, digestive flora, and internal organs. Human bodies do not necessarily want the same food.

Pediatrician Clara Davis began a well-known study in 1928, starting with three infants, Donald, Earl, and Abraham. Over the years, she added twelve more. The children ranged in age from six to eleven months old and were medically or nutritionally challenged in some way. The children had been breastfed with no prior taste associations or desires other than for breast milk. Some had already been offered different foods but had refused them. Thereby began a feeding experiment that is heralded to this day. The children were offered thirty-two different foods at every meal. There were ten different vegetables along with apples, peaches, bananas, pineapple, cornmeal, barley, oatmeal, wheat, chicken, eggs, haddock, sweet and sour milk, raw and cooked beef, brains, liver, kidneys, sweetbreads, and bone marrow. Not exactly baby food. The dishes were unsalted, but sea salt was placed in a separate dish near each child. None of the children ate the same way, nor did they imitate what others chose. Each food was in its own dish and placed in front of the child. Anybody who has fed a small child has observed how they will scrunch their nose and shake their head at the smell of a teaspoon hovering before them or open their mouth if acceptable. "The nurses' orders were to sit quietly by, spoon in hand, and make no motion," according to Davis. Each child wanted a different diet. Their food combinations were odd, as were the mealtimes when certain foods were desired. Earl arrived bowlegged and distressed by rickets from a vitamin D deficiency. He was offered cod-liver oil in addition to the other foods. He drank the strange-smelling oil off and on for three months until he was healed, after which he never wanted cod-liver oil again. The children varied their diet widely, even wildly. Some of them would go on what staff called "jags." Meat jags, milk jags, or egg jags, and then stop. The children's attending pediatrician called them "the finest group of specimens" he had ever seen in children of their age.

Behavior ecologist Fred Provenza wonders why humans prefer eating what is bad for them and avoiding what is good. His studies of herbivores in the wild showed the same traits as the Clara Davis children. Identical species in a single herd forage differently on wildlands according to the innate nutritional wisdom needed for their well-being. Rats with laboratory-induced diabetes will switch to a high-protein diet, if available, and eliminate the symptoms of diabetes. Of course, none of the children in the nutrition study were offered candy, white bread, junk food, or soft drinks. Today, sugary, ultra-processed foods are freely available. And three fourths of the US population is overweight or obese, as are one billion people in the world.

Food manufacturers and "sensory" chemists know what happens on our tongues and palate, the olfactory responses the mouth experiences, and how they affect the brain and our sense of well-being. Many of the flavors people consume today are synthetic, made of esters, ketones, pyrazines, alcohol, and phenolics. Open the jar of strawberry jam in the cupboard, and chances are your nose is greeted by ethyl methylphenylglycidate. Organoleptic scientists study how our sense organs are affected by odor, taste, color, and mouthfeel. Our taste buds were bio-hacked decades ago. We became culinary hamsters in the tread wheel of supermarket food; glistening fat-infested desserts, sugary ketchups, and salty heart-eroding snacks exploit innate appetites that are there to protect us, not kill. Nutritional literacy was reduced to intense tastes: salty, fatty, and sweet. For 99.5 percent of the time humans have lived on the planet, fat, sugar, and salt were difficult to obtain. Until honeybees were domesticated, a wealthy man in Slavic countries was a logger whose axe blade came out of a tree dripping with honey from a wild hive. Salt was scarce inland. Fat came from animals. Eons of scarcity from the distant past hardwire these desires. Big Food knows this and depends upon it. Ultra-processed foods are designed to lure, attract, and addict to make a profit. Their chemists make food ever more desirable: Doritos, Big Macs, and Oreos (there is now an Oreo breakfast cereal for children). Over 70 percent of the American diet is ultra-processed food. These include so-called natural foods such as vegan burgers, protein bars, and oat milk. Consumption of ultra-processed food is directly linked to depression, dementia, diabetes, high blood pressure, stroke, obesity, and cancer. Taste and our extraordinary sense of smell are not playthings to be toyed with. Our senses make *sense* of the world. Junk food and the modern food industry make this intelligence "non-sense."

Food and taste are no longer about the flow of carbon. They are about the flow of money.

When it comes to food, we were predators long before we were consumers. Whether a wild cucumber, clam, or crawfish, we took the food we wanted. The human body is designed to be an expert at obtaining energy from the living world—animal, fungal, or vegetable. Speed, dexterity, teeth, jaws, smell, and hearing have been our bodyguards, enabling us to feed and reproduce. According to Dr. Chris van Tulleken, author of *Ultra-Processed People*, the food industry has flipped the script. We are prey. Our children are prey. Embryos are prey. The food industry is the predator.

Because of a mistranslation of a German paper written in 1901, there is a myth that we taste different flavors on different parts of the tongue. Try it yourself. It isn't true. Each taste bud recognizes the full range of taste. For a hundred years, the world accepted a theory of taste describing how we sensed sweets on the tip of our tongue, bitter foods at the back, and salty and sour to the left and right, even though not a single human being experienced that. That quivering, moist, reptilian projectile in your mouth is a direct extension of millions of years of evolution and learning. Taste and smell are how the body detects goodness or toxins. It is the foremost expression of the immune system deciding what can become you and what should not. When we choose the foods we eat, we either better the world or worsen it, sustain life or dishonor it, improve our health or degrade it. What we eat and how it is grown significantly affects climate and global warming, surpassing all cars, ships, planes, trucks, and railroads in its impact. The food industry severely degrades biodiversity, oceans, rivers, pollinators, grasslands, and animal health.

Taste is sensitive. When we kiss, there is a scent and a flavor. From the body's point of view, it is an avalanche of information. We taste our friend or lover and instantly decide whether to do it again. A raspberry gets that same intimacy. A good meal exchanges precious bodily fluids with herbs, spices, grains, roots, seeds, meats, and oils. When you chew, your tongue and its ten thousand taste buds evaluate hundreds of millions of molecules—sorting, testing, probing, like a doorman making sure the food is on the guest list. Under a microscope, taste buds appear phantasmagorical, like creatures in a Hieronymus Bosch landscape. Fungiform receptors appear to be clusters of toadstools; filiform taste buds look like hooded gangs waving their pointy heads. Foliates run across our palate like desert canals. If there were only a dozen tastes, we would not care much about food. We would be bored, eat cat food, and be done with it. Flavor and the nose create mosaics of experiences we seek with our tongue and mouth. Hold your nose tight, sip a great Bordeaux, and it changes from Cinderella to a crone.

Our noses are the key to taste. Human beings can detect over one trillion olfactory stimuli. The common belief that we have a poor sense of smell is derived from a nineteenth-century theory that has hung around like a bad rash. There is no basis for it. Our olfactory abilities surpass those of dogs and wolves. Canines are sensitive to specific scents that alert them to food, sex, and danger. However, humans have a greater capacity and need to sense the world around them. It has been shown that "super-smellers" can detect the presence of Parkinson's disease in another person from several feet away, and as yet, there is no known biochemical test for it. Our choice of a mate is heavily influenced by scent. Today, our olfactory capacity can be eclipsed by pollution, monotonous foods, and stuffy sinuses caused by allergies.

Take another look at some of the foods first developed in the Americas: cacao, tomatoes, avocados, peppers, chilies, peanuts, cashews, sunflower, safflower, vanilla, pineapple, papaya, blueberries, strawberries, passion fruit, pecans, melons, cucumbers, pumpkins, butternuts, zucchini, cranberries, lima beans, kidney beans, pinto beans.

Do any of these taste the same to you? If you taste one and then randomly another, 33,554,432 combinations of taste are possible. None of us detect scent and taste in the same way, and no taste is repeated exactly in nature, not a pumpkin, pear, or poppy seed. To say we only have five tastes erases food's complexity. What we taste differs, as does each bite of natural food. Industrial food is carefully designed to be uniform, to taste the same the world over. The US Department of Agriculture tracks 150 nutritional components. However, our food has more than 26,000 different biochemicals and phytonutrients.

A public database called the Periodic Table of Food Initiative has been created to remedy the need for more information on food chemistry. Their target is dietary-related disease and death and the agriculture systems that produce impoverished food. Their discoveries are startling. For example, there are nearly 10,000 biochemicals in broccoli, and almost the same in kale. However, the overlap is less than 10 percent. We need to find out what the other 9,000 nutrients do in our body. This new era could transform our understanding of health and nutrition.

Flavor is released by the interaction of water and molecules, in this case, saliva. Saliva contains acids, enzymes, electrolytes, proteins, and cholesterol, which interact with the food and modify its chemistry. Our taste changes when we are in love. Or when we are in fear. If sick with a fever, our saliva may regress to that of a one-year-old when we have little or no amylase, the enzyme required to break down complex carbohydrates. A piece of bread will taste like cardboard. The body focuses on invading bacteria or viruses, not enzymes. At one time, mothers wisely made milk toast for a sick child. Toasting turns the carbs into easily digestible dextrose, and the lactose in milk was our first food. We have cravings when we are deficient in a vitamin or mineral. If a person has sugar cravings, eating plenty of cooked greens seems to lessen or stop it. Some taste buds stay with us for twenty-four hours and then disappear. Others are replaced every week or ten days. After you fast for a week, it is like you never tasted food before, a biological clean sheet for the renewed taste buds. There are superstars of the taste world, chefs who can take one sip of bouillabaisse and tell you when the mullet was caught, the appellation of the vineyard, whether mandarins or navels were used for the orange zest, the variety of tomato, whether the saffron is from Iran or Azerbaijan, if the olive oil was extra virgin or virgin, and whether the olives were arbequina or cornicabra.

Their extraordinary sensibilities pave culinary highways for future travelers. They write recipes like poets write verse, and we go to their food theaters to be enthralled.

Sometimes, we succumb to fads and force ourselves to eat food because we are told it's good for us, even though it may not taste good. Raw and cooked kale could be an example. Recent tests show market kale has unusually high amounts of heavy metals. Raw kale contains goitrogens that suppress thyroid hormones by inhibiting iodine uptake. Farmers avoid feeding kale to cows and sheep because it can cause oxalate poisoning, colic, diarrhea, and anemia. Maybe your initial wrinkled nose was correct.

Mitochondria, primordial enclaves of carbon-based bacteria, power the cells in your body. We feed ancient life-forms that joined us 1.45 billion years ago when we eat. When our authentic taste is regained and experienced, when we have *good* taste, this community of human and nonhuman cells guides us. We cultivate our life, including our second brain in the gut, just as a wise farmer cultivates and feeds the life in the soil, the underground world of plants. The good news is that you are not actually in charge. There is an astonishing energy flow among carbon-based enzymes, blood cells, neurons, wiggly taste buds, antigens, and hundreds of other processes. The flow of carbon is the life you feel and experience. We can name, study, and analyze the body. Still, if we could understand a fraction of what is happening in our body, its complexity and intelligence, we would realize we are in the presence of mystery. Buckminster Fuller once observed that Spaceship Earth is so well designed that we don't know we are on one. Our body is so well designed that many people do not realize they are in one.

A new generation of farmers, chefs, bakers, and makers is providing us with pleasure, restoring our taste buds, and regenerating the land with their unstinting and underpaid labor. Reversing global warming requires changing the food we eat, where it came from, and how it was raised. Restoring pollinators and the soil depends on changing how we farm. In the United States, where 73 percent of the food sold is ultraprocessed and decidedly unhealthy, an inspired movement is reclaiming farmland, locales, climate, culture, and foodways. I will leave it to cultural historians to describe why we let our biology end up in the hands of Tyson Foods and Kraft Heinz. Congressman Earl Blumenauer is concise: "We pay too much to the wrong people, to grow the wrong food, the wrong way in the wrong places." If we are to take back ownership and responsibility for our health and the biological integrity of our land, then we have to take back our mouths and taste buds from those who would use them to accumulate financial capital and return them to those who create our biological capital, away from people who steal the future, to those who heal the future. Let us trust those people who, in Adrienne Rich's words, hold a universe of humility and humus in safekeeping for a world that has a jaded palate, jaded because so much has been lost, a group of people who understand that without our farms, without our exquisite connection to dandelions and thistle honey, dumplings and heirlooms, watersheds and soil, that we will live in a world, in Rich's description, with "no memory, no faithfulness, no purpose for the future, nor honor to the past."

SIX

Sugar Salad

*People are fed by the food industry, which pays
no attention to health, and are treated by the
health industry, which pays no attention to food.*

WENDELL BERRY

W hen I was six months old, I became asthmatic. The
doctors at Mills Hospital in San Mateo, California,
said it was the earliest case they had seen. It was touch and go.
I would gasp, struggle to breathe, and, when I occasionally
turned blue, be rushed to the hospital for oxygen. When I was
fourteen months old, I was strapped down in an oxygen tent for
six weeks with no family visitors allowed. I came out of the tent
no longer knowing who my parents were. They discharged me,
not because my condition had changed but because they needed
the oxygen tent for another patient. Nothing the medical community recommended during childhood came close to a remedy or cure. Later, when doctors started talking to me instead

of my parents, I was told asthma was incurable and genetic, and I would have to learn to live with it. One physician suggested a possible cause could be the relationship with my mother. Another suggested wearing a protective mask and staying indoors with filtered air. Avoiding the outdoors in springtime was the consensus. After numerous scratch tests, it was determined that my body was allergic to over forty common substances. The palliative solution? A prescription drug, aminophylline ephedrine, a stimulant, actually a narcotic, the same ingredient found in ephedra weight-loss formulas that were banned in 2004.

I played sports, and to keep breathing, I ate pills like candy, three times the maximum daily dose. I was stoned on the basketball court for a decade until I read a book by chance. The author was tactless: "If you are sick, it is your fault." What had I done as an infant? I had been listening to the complex lingo of doctors for years, vocabularies that showed me how intelligent they were and how little I knew, and how my "incurable" disease was far more complicated than I could understand. As I continued reading, it became clear that the author was trying to wake me up to not being a victim. My illness was my responsibility and no one else's.

Given the outcome of my medical treatments, I had nothing to lose by setting off on my own. The book recommended a hypoallergenic food fast, food singular. It consisted of rice chewed to the consistency of drool and an infusion made from tea clippings from the *Camellia sinensis* plant. That was it, a gustatory twilight zone with two flavors. On the eighth day, I woke up with an unknown sensation. I felt air deep down in my lungs for the first time. There was no wheezing, no obstruction, no sound. I was nineteen years old. My doctor was not impressed and brushed it off as an anomaly or placebo effect. He wasn't interested in rice and tea. At that time, medical schools did not include a single class in nutrition. Today, twenty-five million Americans have asthma, and to be fair, much of it is caused by air pollution, not innate allergies. However, medical school studies still do not connect food to inflamed airway passages. I was equally ignorant. Why couldn't I eat just like others? Why did the food I prefer cause inflammation? And, for that matter, what was food?

In the following months, I experimented with adding different items to the menu one at a time to experience their effect on my body. If you eat a spartan diet—rice, tea, and vegetables in my case—you immediately feel the difference when you add one additional ingredient: sugar, milk, beer, burgers, bread, coffee, cheese, fries, orange juice, eggs, bacon, tomato, butter, ice cream, chips, and Coke. It wasn't that I didn't want to eat these foods. I was craving them. However, if you eat combinations of these foods daily, which most people do, it is challenging to correlate how you feel, sleep, or think to anything you consume. You might suffer weight gain, athlete's foot, headaches, arthritis, or psoriasis and never be informed that your condition might be related to what you eat and drink. By experiencing food this way, I eliminated one item after another because they didn't make me feel good. I turned to whole grains, seeds (rice, wheat, and oats), nuts, beans, vegetables, fruit, herbs, eggs, and fish. I stopped going to supermarkets and went to farmer's markets. People asked how I could eat such a restrictive diet. However, the standard American diet is narrower than what I practiced. The North American diet consists predominantly of wheat, corn, and rice. Ninety percent of the vegetables consumed are potatoes, tomatoes, onions, lettuce, and carrots. The USDA dietary guidelines take a liberal stance on what constitutes a vegetable: french fries and ketchup qualify as two vegetables. There are thousands of other foods that humans have eaten for millennia, foods Westerners generally do not know, see, touch, grow, or eat. Not only have we domesticated bees, cows, pigs, chickens, and sheep, we seem to have domesticated ourselves.

There is far more nutritional research and knowledge today than in decades past. Yet, the American diet has worsened, the great majority consisting of chemically altered foods that contain unhealthy fats, starches, sugar, salt, and artificial flavorings. Most of what we call food is not food. It is constructed and made of substances that were never part of the food chain: inverted sugars; modified starches; oils that have been refined, bleached, and deodorized into a lubricant; and hydrolyzed protein isolates. The bulk of calories are derived from commodity crops of corn, soy, and wheat. According to Dr. Chris van Tulleken, our foods "have been assembled into concoctions using other molecules that our senses have never been exposed to either: synthetic emulsifiers, low-calorie sweeteners, stabilizing gums, humectants, flavor compounds, dyes, color stabilizers, carbonating agents, firming agents and bulking—and anti-bulking—agents." However, they are more than "constructed," they are designed. There are chemicals in our body today that have never been in the human body for two million years.

If you watch cable news, notice that most advertisements are for pharmaceuticals that treat degenerative diseases: diabetes, cancer, hypertension, stroke, osteoporosis, arthritis, depression, and dementia. Ill health impacts most of American society. Seventy-five percent of young people ages eighteen to twenty-four are unfit for military duty. Forty-two percent of adults are obese. The life expectancy of people in West Virginia is the same as in Syria. In Mississippi, it is lower than in Bangladesh. Adrian Wooldridge points out that a sick America will soon be unable to compete economically with China or defend itself, an outcome of the industrial food system. Beyond the body, industrial agriculture and Big Food are degrading farmland, polluting wells, contaminating rivers, exterminating pollinators, clearing ancient forests, draining wetlands, poisoning farmworkers, and causing irreparable harm to children's futures. In essence, we are eating exploitation, and what we are eating is addictive.

The processed food industry doesn't see it that way. I once met the chief sustainability officer for McDonald's at his request. He came to my office, and I showed him five different ingredient listings and asked if he could identify the foods. He was puzzled by the quiz but gave it a go and finally confessed he had no clue. I explained that each was on the menu at McDonald's and could he try again. He studied the lists but shook his head. I then showed four nutritional panels, the FDA-mandated analysis showing grams of carbohydrates, protein, sugar, fat, fiber, sodium, cholesterol, and total calories. I asked, "Which was the double cheeseburger, and which was the McDonald's salad?" He stared for a minute and claimed it was a trick question. I asked why he thought it was a trick. "Because the highest-calorie food is probably the salad, not the cheeseburger." Bingo. It was the fourteen hundred calorie "sugar" salad. McDonald's ditched the salad a few years later to "slim down" on their costs.

Human existence is a flow of carbon, from food to breath, the energy rhythm captured and released by carbon's gregarious chemistry. Photons create sugar in the fluttering leaves of plants that feed humans, animals, insects, fungi, microbes, and soil. The body receives a daily flow of nutrients that energize our cells. When we break down food into components—glucose, polysaccharides, creatine, casein, glycine, omega-3 fats, vitamin K, etc.—we are lost. The nutritional understanding of essential food components is correct. But it is not nutrition if we conceptually aggregate dozens of components into a healthy diet using supplements, green drinks, or food additives. The addictive approach to taste and mouthfeel has produced a monstrous industry for people struggling with obesity and metabolic disease. One could look at it as collusion between the processed food industry and the pharmaceutical industry. The weight loss industry is estimated to be $377 billion worldwide. Amazon sells over sixty thousand different diet books. The supplement industry is over $150 billion. This is not a flow of carbon. It is a tangle of confusion. People who are overweight feel blamed if not shamed. Consumers live in a food system designed to cause overconsumption. Our diet has changed more in the last 140 years than in the previous one million years. As Michael Pollan succinctly stated, our food became "foodlike."

The United States spends $4.5 trillion annually on a broken healthcare system devoted to disease, 20 percent of total economic activity. My asthma was not cured by rice and tea. The body healed because of what I didn't eat. Eric Roston describes how humans inflict suffering upon themselves and other forms of life; he writes, "Today, we too often live as if humanity—or our country, or we ourselves—is the center of everything. We can do whatever we want without concern for nature's consequences." The fast-food industry spends more than $5 billion a year convincing youth and children to feed addictive desires, not their bodies. The day will come when this will be seen for what it is: a crime against humanity.

The living world puts molecules together that want to be together. This is called green chemistry, a term coined by John Warner and Paul Anastas, the planet's chemistry since life began. Nourishing food does not require force, additives, or food chemists. Over nine thousand years, farmers in the Americas crossed and bred several thousand cultivars of corn. The corn was wildly divergent and significantly nutritious. Today, corn is grown in sprawling monocultures on farms larger than small cities. Over 90 percent of the corn produced is genetically modified to resist herbicides. Its uses include feed for pigs and cows, inexpensive sweeteners for soft drinks, fuel for cars (ethanol), feedstock for plastic, and processed starch for corn chips.

The Mayan culture was one of the most sophisticated

civilizations in the world, with an enlightened knowledge of nutrition. The average Mexican drinks 487 cans of Coca-Cola a year, a number that has doubled in the past ten years. One in six Mexicans has diabetes, the country's number-one cause of death. The sixty-second president of Mexico was the president of Coca-Cola Mexico. Diverse foods and localized varieties created the intricate culture of Mexico. Stripped of its native foods, the country was flooded with industrially farmed, low-cost, sterile corn, due to the North American Free Trade Agreement (NAFTA). Two million traditional corn farmers went bankrupt.

Dr. Weston Price was a Cleveland dentist who traveled the world in the 1930s to search for people not plagued with tooth decay and disease. Price noted in the preface to his classic work, *Nutrition and Physical Degeneration*, that his entire medical education focused on pathology. To discover what created physical well-being, he journeyed to isolated villages in Switzerland and the Outer Hebrides, First Peoples in Arctic Canada, tribes in Africa, Mayan descendants in South America, Aborigines in Australia, and Māori in New Zealand. His exploration revealed how native cultures, when they switched from traditional to processed foods, developed dental caries, narrowed facial and pelvic structures, crooked teeth, and chronic disease. He took samples of traditional foods in each location and analyzed their nutritional profiles, showing them to be four to ten times higher in essential minerals and vitamins than American fare. His wife took photographs wherever they went that showed the healthy faces and bone structure of the women and men who ate traditionally. These were in stark contrast to the physiognomy of children whose parents ate Western processed foods.

Indigenous people in the Americas began to cook corn in an alkaline solution of water and lime ten thousand years ago, a process called nixtamalization. The alkaline preparation of the corn releases bound-up niacin and calcium while softening the corn so that it can be made into masa, the basis of tortillas, tostadas, and tamales. Without this processing, corn-dependent cultures would suffer from pellagra, a disease marked by diarrhea, dementia, and dermatitis. How did ancient cultures prevent a deficiency without knowing that the cause was the lack of niacin? There had to be some serious research and development. People tested, sampled, tasted, cooked, roasted, dried, and fermented local foods in myriad manifestations over thousands of years. These cultures were experimenting and refining their diets when Paris, London, and Berlin did not exist.

Historian John Mohawk, a Seneca born into the Turtle clan, describes pre-Columbian Turtle Island (America and Canada) as a continent where hundreds of tribes and nations resided in bioregions, learning how to live and thrive with regional plants and animals over hundreds of generations. That Turtle Island is referred to as the New World speaks to the pervasive settler delusion. According to Mohawk, these cultures were not based on money, and food was never sold. Mohawk described a culture where "everything that ever happens to you is watched. When you're small, if you don't thrive, they notice. If they feed you something and you don't thrive, they notice. If they feed you something else and you do thrive, they notice. Every possibility they have at their fingertips can be tried; they are motivated to watch and see which foods help people the most. Not which foods help people make money, but which foods have the best biological impact, especially on young and old people."

The Seneca live in western and central New York, where there are other tribes and clans. They were gardeners, farmers, and horticulturists. Mohawk remembers at least twenty varieties of corn and dozens of types of squash, beans, and greens. The greens they ate were wild harvested in the woods and meadows. Around them were blackberry, blueberry, currant, staghorn sumac, elderberry, wild apple, black cherry, fruits, butternuts, chestnuts, shagbark hickories, black walnuts, pecans, hazelnuts, acorns, chicken of the woods, chanterelles, hen of the woods, lion's mane, puffballs, wild rose, dandelion, mayapple, wild leeks, domestic fowl, turkey, brook trout, sturgeon, shovelnose, walleye, white and yellow bass, pigeons, squab, deer, and elk. "The only benefit they [the Seneca] were interested in was the people's health. The only one. The health of the little people, middle-size people, old people, but always they're thinking about the health." For tribal cultures in the world, the task was always straightforward. When a society is responsible for its health, it will be vigilant about what it eats and consumes, especially when food is seen as deserving of reverence, respect, and gratitude.

For hundreds of thousands of years, there was an ongoing process of plant selection by *Homo sapiens*: which were food, which were medicine, and which were toxic. North American medicinal plants were long known to Native Americans, from bear root and echinacea to goldenseal and American ginseng. Colonists did not respect the breadth of their botanical intelligence.

Because of disease, brutality, and deracination, much native knowledge was destroyed and may be lost. The aboriginal population in Australia before the arrival of colonists is thought to number from 750,000 to 1,500,000. One hundred years later, it was closer to 100,000. Nevertheless, ecological knowledge is being recovered. In Australia, ethnobotanist Beth Gott collaborates with Koori people to create an Aboriginal garden at Monash University, growing 150 edible plants, from soporific dune thistles to hearty daisy yam. She and her students cataloged more than 1,000 species that were a part of Aboriginal "bush tucker."

Where I live in Northern California, the local Miwok ate serviceberries, barberries, strawberries, toyon berries, cream bush berries, plums, cherries, gooseberries, rose hips, blue elderberries, snowberries, huckleberries, and grapes. For nuts, they chose hazelnuts, walnuts, and pine nuts. Roots included onions, ginger, cluster lilies, native tulips, camas, fritillaries, desert parsley, evening primrose, and cattails. Vegetable greens were monkey flowers, sage, clovers, violets, mule ears, redbud, and cottonwood flowers. Wild peas and grasses were harvested for their seeds. Tea was lilacs, self-heal, and fir, sweetened with maple syrup from bigleaf maples. From the rivers, they caught salmon, trout, and lampreys. No obesity, no asthma, no heart disease, no dementia, no Alzheimer's, no psoriasis, no type 1 or 2 diabetes. None of us can eat that way anymore. But we can make choices. Over 160 countries in the world have chosen to ban American bread, corn, candy, beef, pork, and other foodstuffs because they contain ingredients considered toxic.

Perhaps the most compelling and eloquent book on sustenance is *Nourishment* by Fred Provenza. His scientific career involved animals, wild and domesticated. As the Clara Davis feeding study demonstrated, animals intrinsically know what to eat. Or, in the case of the human animal, did know. We are overwhelmed by hucksters, advertising, corporations, academics, and myths. Provenza writes, "Nobody has to tell a wild plant, bacteria, insect, fish, bird, or mammal what to eat to sustain health, how to self-medicate to recover from disease, or how to develop and reproduce. Ironically, people now must be told by 'authorities' what to and what not to eat. Do humans lack the ability to identify and choose nourishing foods or has that ability been hijacked?" The climate crisis is not up in the sky. It is down here, on dinner plates, in take-out cartons, drive-up windows, denatured soils, and confined-area feeding operations for cattle, chickens, and pigs. The crisis is a direct outcome of what we choose for food.

SEVEN

Bucky and Bing

God is a verb, not a noun.

BUCKMINSTER FULLER

Spaceship Earth was a metaphor that architect and engineer Richard Buckminster Fuller used to guide human activity. A long-distance voyage in space would require maintenance, cooperation, teamwork, fairness, and a profound understanding of life-support systems. The planetary spaceship we are on is ingeniously designed. Passengers don't realize they are flying a million miles per hour through space without seat belts, lots of room in coach, and delicious food. The spaceship came with operating instructions and essential guidelines, including everyone is crew; don't poison the water, soil, or air; and be sure the spaceship doesn't get too crowded with passengers. And an important rule: Don't touch the thermostat.

I used Buckminster Fuller's metaphor to lead a workshop for the management of a world-famous chemical company that took pride in its thousands of products, particularly its highly profitable arsenal of chemical herbicides and pesticides. The group was divided into five teams, and the task was to spend the day designing a spaceship that would support their lives and future generations for one hundred years before returning to Earth. It could be as large as they wanted and receive light, as long as it did not release any waste into space. End of the day, the teams presented their proposed designs to each other. They then voted for the spaceship they would choose to journey on. There was a hands-down winner. Instead of taking endless amounts of digital entertainment, the winning spaceship, called *Genesis*, populated itself with artists, singers, dancers, playwrights, actors, and poets. The crew and passengers included many nationalities, traditions, and ethnicities. The other four teams focused primarily on science and did not mention art, tradition, or diversity.

The *Genesis* spaceship population was chosen with the hope that it could develop a one-hundred-year culture. That required fair and equitable distribution of resources. *Genesis* decided not to take any products their employer manufactured. None could be recycled and were too toxic for a closed system. When explicitly asked about pesticides, the company's cash cow, the team explained that insects are essential to maintaining a healthy ecosystem. Herbicides? Not those either because "weeds" feed soil fertility and pollinators. The winning team started an organic garden on the corporate grounds soon after, and three members quit. On today's Earth spaceship, 1 percent of the people own and control nearly half of all planetary wealth, resources, energy, food, and land. Half of the population controls 1 percent of land and resources.

Buckminster Fuller's most famous design, the geodesic dome, was inspired by water. When he was an ensign on a naval crash boat, Fuller stood at the stern looking at the wake, wondering why bubbles were round. It turns out that a "bubble" can withstand the greatest load and contain more space, employing less material than any other structure. A geodesic dome could be built quickly with prefabricated components, and there was 30 percent less surface area than a conventional building, which reduced energy heating and cooling. Tens of thousands were built worldwide as housing, greenhouses, theaters, biomes for the Eden Project in the UK, wind-resistant radar stations in the Antarctic, the Salvador Dalí museum in Spain, and an eco-hotel in Chilean Patagonia. The largest geodesic dome, 708 feet in diameter, encloses the entire stadium for Japan's Fukuoka Hawks baseball team.

In Fuller's lifetime, there were three known forms of pure carbon—graphite, diamonds, and amorphous carbon, known as soot or charcoal. When you touch the surface of graphite, you're stroking a million stacked, silky-smooth sheets of pure carbon one atom thick, each layer a hexagonal array of atoms. The carbon in a diamond is organized into three-dimensional crystalline structures. In 1985, scientists discovered a fourth configuration of carbon that engendered the field of nanotechnology, particles that measure less than one billionth of a meter in size. The goal of nanotechnology is to manipulate materials at an atomic scale. Imagine a transistor the size of a molecule.

The discovery occurred when six scientists, led by Harold Kroto, Richard Smalley, and Robert Curl, studied molecular carbon chains using optical spectroscopes, which analyze light intensity over the electromagnetic spectrum. Prisms suspended in a window divide the sun's white light into a rainbow of colors. A spectroscope is similar. It splits incoming radiation and can determine an astronomical object's composition, density, and temperature by analyzing the wavelengths. Each molecule has a signature frequency; one could even say a unique sound if it were audible. The scientists focused their spectrometers on interstellar gas clouds created by dying stars and found unknown chains of carbon molecules. Curl suggested they try to replicate the conditions they were measuring light-years away by using a pulse laser at Rice University in Texas. The laser vaporized graphite atoms at similar temperatures found in dying red stars.

The graphite was converted into plasma, a hot ionized gas that strips electrons away from atoms. Fifty miles above, the Earth's atmosphere changes from a gas to plasma, sometimes seen when the charged particles from incoming solar radiation create rippling curtains of fluorescence known as the aurora borealis. Though it is rare on Earth, 99 percent of the universe is made of plasma. It is the mother ship of the other three forms of matter—gas, liquid, and solid. As the ionized plasma cools, it returns to a solid state where carbon atoms organize and combine. Using their spectrometers, the scientists found a profusion of carbon molecules containing sixty carbon atoms. Known as C60, such a molecule had never before been imagined. How was it structured? What did it look like? The scientists were puzzled about how sixty carbon atoms could organize themselves into a stable macromolecule. Smalley played around with scissors and tape to make dome-like sixty-carbon spheres. They knew carbon makes hexagons and pentagons, so Kroto suggested combining both forms because he had once made a Christmas star for his children in this way. It solved the puzzle—a spherical cage with thirty-two faces, twelve pentagons, and twenty hexagons. It was a hollow molecule, the most symmetrical and aesthetically beautiful molecule known. Physicists named the new configuration a fullerene, or buckyball, because of its similarity to Buckminster Fuller's geodesic domes.

The 1985 discovery riveted chemists around the world. It was like discovering a planet hiding behind Jupiter. Carbon is by far the most studied, analyzed, and researched element in physics and chemistry because it is the basis of life and most materials found on Earth. After that event in 1985, chemistry textbooks had to be thrown out. There were not three basic structures of carbon; there were four. Smalley called the breakthrough a chemist's Christmas. He compared its importance to the 1825 discovery of the six-carbon benzene ring, a toxic hydrocarbon that became the basis for most of the synthetic chemicals made today. And as predicted, fullerenes set off a cascade of inventions and potential uses. The spherical structure does not dissolve in water and can roam around the body to deliver controlled releases of drugs to specific sites. Fullerenes can be used for gene delivery, placing and splicing foreign DNA into particular groups of cells. As an antiviral agent, it transports protease enzymes that stop the replication of the human immunodeficiency virus (HIV), which in turn delays the onset of AIDS. Fullerene derivatives inhibit the hepatitis C virus. A trademarked water-soluble form of a buckyball is called a "radical sponge," which is highly effective in protecting the skin from the free radicals emitted by UV irradiation and is now included in some over-the-counter sunscreens. And move over blueberries and turmeric—fullerenes are considered the most potent antioxidants in the world. A 2012 study showed a near doubling of the lifespan of rats fed C60-containing olive oil, presumably due to reduced oxidative stress associated with old age (the test has not been repeated).

Waves of research revealed variations of fullerenes, containing from 28 to 108 atoms. In 1991, a Japanese scientist discovered nanotubes—elongated carbon sheets arranged in tubular form, capped on both ends like a gelatin capsule, a nanometer in diameter. Human hair averages eighty thousand nanometers in width. Structurally, nanotubes are one hundred times stronger than steel, with one sixth the weight. Today, nanotubes are being made on a massive scale. It is a multibillion-dollar industry offering conductive, strengthening, and lightweight properties to various materials, including glass, composites, sensors, semiconductors, aluminum, paints, and ceramics. Dozens of industries employ nanotubes, including aerospace, biomedicine, electronics, wind turbines, batteries, and solar. Nanotube integration could reduce electric vehicle weight by up to 25 percent for cars or scooters, granting a 30 percent increase in overall efficiency. They not only create lighter car bodies, but they also reduce rubber tires' rolling resistance. Nanotubes became a revolutionary, once-in-a-generation technology.

However, nano comes with a no-no. Nanotubes being manufactured and sold are rarely pure carbon. They contain catalytic metal coatings, including nickel, cobalt, and molybdenum. There are now over fifty thousand configurations of nanotubes. They have different dimensions, properties, and added compounds. Their light weight allows them to be invisibly and unknowingly present in the air. They have no taste, color, smell, or tactile sensation. Nanotubes slough off the composite materials they are intended to strengthen, similar to asbestos fibers, with potentially a similar outcome to human lungs. They are more persistent than any pesticide, insoluble in water, and do not biodegrade. Using, cleaning, and disposing of engineered nanotubes can damage human health and pollute the environment. For workers, inhalation and exposure can cause high blood pressure, emphysema, heart attacks, and kidney damage. Ingestion can cause cancer, DNA damage, significant inflammation, weakening of the mitochondrial membrane, and accelerated cell death. The pharmaceutical industry values nanotube technology because it can penetrate the skin and even a single cell. That means precise drug delivery to pathological sites with fewer side effects. It also means that forms of nanotubes in the environment can enter the body through contact, indicating that nanotubes can insert metallic or polluting compounds directly into the skin and lungs.

The ubiquity and bioavailability of nanotubes mean they will move up the food chain in the same way as DDT and glyphosate. Manufacturers clean production facilities with water and acids flushed into sewage systems. It is one thing to have known toxins, such as a quart of paint thinner in the garage. It is another to have ubiquitous, immeasurable, long-lived toxins everywhere. The world's largest manufacturer of nanotubes is a Chinese/Russian joint venture; neither country is known for vigorous environmental protection or regulations. It is unlikely that doubts, concerns, or reported dangers will slow the explosion of nanotechnology. The bullishness of science is best expressed by Mihail Roco, senior advisor for nanotechnology at the National Science Foundation: "We have about 100 kinds of atoms, and right now 20–25 are frequently used. One should be able to use all of them in various arrangements at the nanoscale, exploiting their properties however we like." This is prototypical of the cloistered, dissociative beliefs of Western science.

Would Buckminster Fuller want to see the commercialization and distribution of fullerenes, buckyballs, and nanotubes in the Spaceship? Those questions tend to be quickly dismissed with assurances that safety precautions will be implemented. The enthusiasm for nanomaterials in scientific and engineering communities is giddy and expansive. A "nano car" was built with four buckyballs for the wheels and a few nanotubes for the frame. It had no steering wheel, which might unintentionally symbolize the nanotechnology world. The word commonly used in nanotechnology is that science can, for the first time, "domesticate" atoms. The temptations are overwhelming. Scientists even speak about merging nanotechnology into the human cell, like an artificial organelle similar to mitochondria.

The term *domesticate* means to tame, master, and subdue. There is a precedent—science domesticated molecules employing organic chemistry (the study of carbon-containing compounds) starting in 1828. The first synthetic product was urea, which is still used in fertilizers, drugs, and plastics. The scale of organic chemistry now encompasses over 350,000 synthetic chemicals and chemical mixtures. It is estimated that 220 billion tons of chemically active materials are released into the environment annually, discharged by industrial agriculture, fossil fuel mining, oil refining, construction, pharmaceutical companies, and manufacturers. The composition of more than fifty thousand chemicals remains confidential and not revealed to the public or regulators.

In most cases, released chemicals build up cumulatively over the years. More than five hundred reported dead zones in the world's lakes and oceans are caused by agrochemical runoff. Carcinogens, flame retardants, PFOAs, PCBs, heavy metal compounds, endocrine disruptors, phthalates, and glyphosate are found in most people alive today. These chemicals were introduced as having significant benefits to humankind when first manufactured and sold into the marketplace. The safety assurances concerning nanotubes are being proven false; government regulation cannot keep up with existing chemicals. For example, the EU bans phthalates in flooring material, not food packaging. In the United States, they can be used in lipstick but not sippy cups. According to *The Guardian*, "researchers have linked phthalates to asthma, attention-deficit hyperactivity disorder, breast cancer, obesity and type II diabetes, low IQ, neurodevelopmental issues, behavioral issues, autism spectrum disorders, altered reproductive development and male fertility issues." Add to the regulator workload approximately fifty thousand forms of nanotubes.

Annual fullerene production is relatively small in pounds but not in number. Approximately six thousand tons of the yearly output equals three nonillion fullerenes. What is a nonillion? Place thirty zeroes after the three. In 2018, Drs. Rasel Das, Bey Fen Leo, and Finbarr Murphy analyzed the literature examining the uncertainties and risks involved with fullerenes concerning one of the areas in which nanotubes are being applied: water purification. Nanotubes are unmatched in removing chemical and biological pollution because of their relatively large surface area and reactivity to chemicals. The problem is that filters release some of the nanotubes into fresh and ocean waters. They cannot be disposed of in landfills or waste incineration plants. They don't degrade; they do not dissolve; they are extraordinarily strong—the qualities that make them valuable make them permanent new inhabitants of the environment. Just as with charcoal, coal, oil, and gas, the allure of carbon for humanity is overwhelming. The fact that carbon collaborates so easily is its compelling attraction. Do we know what we are doing with carbon? Or, to put it another way, are we collaborating or coercing? Fuller was fond of saying when he was working on a problem, "If the solution is not beautiful, I know it is wrong."

In the past ten years, a novel carbon technology emerged from the study of nanotechnology that is quite different. It was invented by Liangbing (Bing) Hu. Born on a rice and cotton farm in Hubei Province, Bing was chosen as a student for the School of the Gifted Young and placed in university at fifteen. He focused on physics, and his boyish fascination then and now is the capacity of physics to study matter at any scale. Materials from galaxies to subatomic particles could be researched, explored, and examined. Bing received a doctorate in nanotechnology at UCLA when he was twenty and went on to do his postdoc in materials science and engineering at Stanford.

When one speaks and listens to Bing, there is something similar to his way of seeing the world to that of Buckminster Fuller. In my short time with Fuller, the best way I could describe his mind is brilliant innocence. Bing looked at wood the way Buckminster Fuller looked at bubbles. Why was cellulose so strong? He went small to find out. When studied with an electron microscope, the cellulosic fibers in wood were structurally similar to nanotubes, growing spirally upward in one direction. It turns out that cellulosic fibers are stronger than carbon fiber. The first time we met, Bing demonstrated the strength of cellulosic fibers. He effortlessly tore half a sheet of office paper, as could any child. He got a new sheet, held it from both ends, and tried to pull it apart. You and I cannot. Cellulosic fibers are simple: long chains of glucose monomers, a molecule that can be

endlessly bonded to itself, precisely the same as carbon nanotubes, but ten thousand times less expensive.

Bing believed that if you could utilize the strength of cellulosic
nanofibers in a piece of engineered wood, it would create a new
material, and its properties would rival steel. At his University
of Maryland laboratory, Bing and his associates began to experiment. Wood was boiled with sodium hydroxide to break
down the lignin. Boiled again to remove the chemical, the
wood is hot pressed at 212 degrees Fahrenheit to one fifth the
thickness. The boiling opens up spaces inside the wood, allowing compression to bind the hydrogen atoms in the cellulosic
fibers to each other. The result is what he calls InventWood, 50
percent stronger than steel, one sixth the weight, and one half
the cost. It is functionally fireproof (too dense), and insects
can't munch on it due to its hardness. If honeycombed, it can be
sandwiched between InventWood panels to make floors as
strong and soundproof as concrete. Imagine a conventional
fifty-story high-rise building 180 square feet at the base. The
concrete and steel required to construct it would weigh approximately 250,000 tons, with 12,500 truckloads to the job site. If
built of InventWood, it would come in at one twentieth the
weight. In a conventional steel and concrete building, the structure on the first floor supports forty-nine floors of steel and
concrete above. If the Twin Towers at the World Trade Center
had been built of InventWood, neither would have collapsed.
The towers pancaked from their sheer weight once the steel in
one floor was weakened by burning fuel from the aircraft.

Bing's invention can be shaped and molded into different

configurations using a hot press. A Boeing 787 Dreamliner or Ford F-150 truck built of InventWood would be lighter, cheaper, and safer. A thin panel of InventWood will resist a bullet. If infused with methyl methacrylate, the wood becomes translucent and can replace plastic or glass in structural applications. The difference between steel and concrete goes further. Steel is responsible for approximately 8 percent of global greenhouse gas emissions, not counting the impact of transporting it to locations where it is needed. InventWood does the opposite; it sequesters carbon. It can be made of different types of trees, with the ideal substrate being bamboo, which is a grass. To replace half the worldwide production of steel, utilizing bamboo as the sole material source, about twenty-five million acres would be devoted to growing bamboo, just over 50 percent of Nebraska farmland, or less than .001 percent of the farmland in the world. It would sequester 106 million tons of carbon annually, replacing 1.7 billion tons of annual emissions from steel.

The above are estimates. What is not speculative is the technology, its affordability, strength, and practicality. How quickly it is taken up and utilized is to be determined. We see imagination and practical genius here, much like what people saw in Buckminster Fuller. The US Department of Energy considers it genius, too. In November 2022, it awarded Bing's company $20 million to build a pilot factory in Maryland. What Bing imagined and perfected over ten years of research is a different carbon world, a cellulosic age that reduces human impact upon the environment by a factor of more than three thousand times by understanding the plant world in an entirely different way.

Bing is three years into a project to exploit the nature of cellulosic nanofibers for batteries. Instead of compressing the fibers and making hydrogen bonds, the fibers are separated. Imagine the finest filigrees of minute invisible strands of hair that would allow the free movement of lithium ions. You would have a lighter battery encased in InventWood.

EIGHT

Green Beings

Nature needs no home; it is home.

DAVID GEORGE HASKELL

If you were to examine a new plant daily, it would take 1,200 years to check out all of the species on Earth. You would eventually meet the twenty-pound rafflesia flower that is fifteen feet in circumference, the giant Bolivian water lily upon which a seven-foot basketball player could easily stretch out on and nap, and the Tasmanian King's Holly still attached to its 135,000-year-old base clone. A high point might be the elephantine baobab tree that can measure eighty-seven feet in diameter (imagine a tree as wide as an eight-story building lying on its side), trees so voluminous they have been hollowed out for prisons, pubs, and warehouses. The Mediterranean prickly oak can be burnt to the ground, grazed into stubble, felled by axe, and regrow to support munching goats standing on its uplifted branches. The leguminous dynamite tree (*Hura crepitans*) explodes its seedpods with firecracker intensity, ejecting seeds three hundred feet into the air. A seagrass colony discovered off the coast of Ibiza is nearly 200,000 years old. Plants are a measure of biodiversity, but you rarely see the plant kingdom discussed in the context of climate on the same level as wind turbines, solar farms, battery storage, green buildings, and electric vehicles. Yet, grasslands, forests, seagrass meadows, shrubs, mosses, and vines comprise the most significant flow of carbon on the planet, ten times greater than fossil fuel emissions and other human activity combined.

Plants and animals have a common origin: eukaryotic cells that contain a nucleus and a skeletal membrane. Eight hundred million years ago, one branch of eukaryotes' cells became animals, another plant, and a third fungi. The core functional difference between plants and animals is movement. Plants are stationary—they are sessile. Animals move in air, water, and land—they are mobile. Plants cannot run if threatened. The plant world was long regarded as inanimate. Plants don't howl or chirp, aren't charismatic, have no personality, and are fixed to a spot. Despite the presumed advantage of being an animal, or perhaps because of the disadvantage of being stationary, plants developed twenty senses that perceive and respond to their environment compared to the five senses possessed by animals. They don't need to go anywhere to eat because they make food from light, air, soil, and water. They have great sex remotely courtesy of pollinators and the wind. They have modular, decentralized vital functions, whereas animals perish when their organs are attacked. No one part of a plant is essential. If gnawed to the ground, most can regenerate. Grazing and chomping can create a more vigorous plant, which is why we prune. You would have no success sowing a claw, tail, or ear, but a new plant can be created using a cutting, leaf, or root. We are individuals, which means "not divisible." Plants comprise colonies and systems that thrive because they are divisible. That doesn't mean they can't touch, talk, taste, hear, or smell. They can.

Weddings, bat mitzvahs, quinceañeras, and funerals would be incomplete without the symbolic presence of lilies, roses, peonies, and daisies. Throughout human history, plants have been revered. Trees were worshipped, tobacco was sacred, sweetgrass was hallowed, holy basil is its name, peyote was venerated, and maize, squash, and climbing beans planted together were exalted as the Three Sisters. Teachings, myths, and parables about the life and meaning of plants ebbed in the twentieth century, possibly due to the advent of plant biology. Plant science heralded an extraordinary exploration of the plant world, resulting in new vocabularies such as *cryptobiosis*, *phytophilous*, and *diplostemonous*.

In 1913, German scientist Fritz Haber was tinkering with nitrogen to make better mustard gas weapons for war. He incidentally created a method to split atmospheric nitrogen into synthetic nitrate. Following that discovery, Carl Bosch worked with Haber to develop a high-pressure method that could make vast quantities of affordable ammonia fertilizer. For the first time, soluble nitrogen could be added to the top layer of the soil. Using the Haber-Bosch process, yields doubled and tripled. Soil gradually regressed to a chemical medium that held up crops. Beliefs that plants were sacred, to be revered and honored, were pushed aside in favor of their capacity to create wealth. Plants could be manipulated just like any other industrial material. Before and after the advent of inexpensive fertilizers, groundbreaking research was conducted by two agricultural scientists, George Washington Carver and Luther Burbank. Carver focused on moving away from the cotton monocultures of Mississippi and Alabama and restoring diversity and soil fertility to improve plant and human health. Burbank was a horticultural magician who created eight hundred varieties and strains of fruits, nuts, vegetables, trees, and flowers by repeated crossbreeding at his farm in Sebastopol, California.

Burbank would plant thousands of seedlings of a single variety, looking for variants that improved yield, taste, color, or fragrance. Once selected, desirable variants would be sown and cross-pollinated, which he meticulously did by hand. Results included delectable Burbank plums, the Flaming Gold nectarine, freestone peaches, and chestnut trees that yielded crops in three years instead of twenty-five. He created a blight-resistant potato to prevent another Irish famine. Today, the Russet Burbank is used in McDonald's french fries. Burbank wrote an eight-volume series about his work entitled *How Plants Are Trained to Work for Man* that foretold the future of agronomy.

The modification of plants is so prevalent today that we hardly notice. Plants are for food, decoration, fiber, and timber. Monsanto developed glyphosate-resistant corn and soy under the Burbank rubric, retooling the plant's genetics to serve "mankind." Their actual purpose for reengineering commodity crops was to serve Monsanto. Their herbicide glyphosate had gone out of patent, and profit margins were crashing. Monsanto genetically modified corn and soy to be compatible with their chemically altered glyphosate. It was the first time an herbicide did not kill the crop, just the competing plants called weeds. The genetically modified seeds and herbicide were inseparable. A farmer couldn't buy the patented seeds; they could only purchase a license as if it were software. The license agreement prohibited growers from following the ancient practice of replanting seeds from their crops. Farmers found replanting the modified seeds were sued, and many were forced to pay damages. It was a brilliant and devastating financial ploy. Today, glyphosate is the world's leading herbicide, with approximately 1.5 billion pounds used yearly. Because it is a water-soluble pesticide, it travels throughout the land and is found in dairy cows, house dust, drinking water, ice cream, tampons, organic cereals, sea lions, breast milk, and 75 percent of the world's rain.

Christmas trees, once used to celebrate the birth of Jesus, are tossed onto the streets as trash after the holidays. South American drug cartels produce pesticide-drenched roses for (Saint) Valentine's Day to launder money. Scientists promote the idea of plants and trees being "weaponized" to combat climate change. Farm-grown biofuel will replace refined kerosene to power aircraft. Genetically modified soy will be bio-plastic feedstock, and cultured meat will be grown in towering stainless-steel tanks. In the United States, 107 processing facilities dry and process wood into eleven million tons of pellets per year for the "renewable" energy industry—the astonishing proposition that burning trees addresses climate change favorably. According to author and plant raconteur Richard Mabey, plants "have largely been reduced to the status of useful and decorative objects. . . . They have come to be seen as the furniture on the planet, necessary, useful, attractive, but 'just there,' passively vegetating. They are certainly not regarded as beings in the sense that animals are."

However, they are most definitely beings, and highly pragmatic ones at that.

Plants have developed remarkable methods to know what is happening around them. Bacteria contain about one thousand genes, and fungi ten thousand. Humans have about the same number as mice, twenty-five thousand genes. Flowering plants can have up to four hundred thousand gene expressions. That doesn't make plants more intelligent but underscores the cumulative complexity that evolved over 470 million years. Because they are immobile, plants sense their surroundings in novel ways. Leaves, stems, and needles are covered with sophisticated networks of interconnected photoreceptors that adjust to capture different wavelengths of light. Animals have four types of photoreceptors in their eyes; plants have thirteen, including one that detects UV light. Plants interpret the amount, color, and direction of light to regulate physiology, growth, and development. Stomata on the epidermis of leaves open with the dawn chorus and begin a daylong exchange of oxygen for carbon dioxide. If it is too hot during the day, stomata close to prevent water evaporation, reopen when cooler, and then contract again at night. This is hardly different from humans, who rise to light and sleep when dark. When attacked by insects, plants emit volatile compounds that other plants detect by smell and taste. Plants have a hygrometer—the ability of roots to detect and precisely measure soil moisture levels. They respond to tactile cues, allowing roots and vines to wend through obstacles. Plant roots dowse for distant sources of water and direct their growth accordingly. They have a perfect nose for belowground nutrients, including nitrogen, phosphorus, calcium, and trace minerals. Look up at the canopy in the rainforest, and you will see how higher trees of the same species share the light by carefully refraining from overlapping. Plants donate 30 percent of their sugars to the soil below, nourishing the community of bacteria, protozoa, fungi, ants, earthworms, and insects that feed on them. When a specific mineral is needed, plants send specific chemical signals to fungi and bacteria that release minerals from sand or rock enzymatically.

After publishing his heavily contested theory of evolution in *The Descent of Man* and *On the Origin of Species*, Charles Darwin focused on plants, soil, and worms for the second half of his life. Darwin and his son Francis began a series of experiments in the 1850s involving plant movement. Their work and conclusions were also scorned by the scientific establishment, criticisms that later proved incorrect. They looked at plant behavior using a timetable befitting an organism that can't move. Plants slowly sway, swivel, stretch, and move with obvious purpose, visible to a patient observer. Rather than plant movement being random, the observant Darwin saw it as purposeful, meaning there must be cause and effect, a call and response. Darwin proposed that a control center resided in sensory root tips "like the brain of a lower animal." Beneath the soil is a world that is more complex than what is above. If you pull up a plant as slowly and gently as possible, the number of roots in hand is less than one thousandth of those left in the ground below. Botanist Howard Dittmer, working with a team of his students, counted 14,335,568 root and root hairs from a single winter rye plant. The total surface area of belowground roots was 130 times greater than the leaves and stems aboveground. Plausibly, the giant dipterocarp trees of Borneo might have several hundred million roots, possibly a billion. Invisible threads in the soil form a plant's sensory organ for which we have no precise word. Root tips discern hard and soft obstacles, detect and avoid pollution, and connect to the hyphae of mycelia networks, where they conduct an astonishing number of transactions, bartering their carbon in exchange for minerals and nutrients. How plant roots process the information received above- and belowground is not understood. Unlike animals, no brain, neurons, nervous system, or physical locus of intelligence exists. Plant neurobiologist Stephen Mancuso believes that descriptions of how the brain works are just as valid for a plant. "The neuron is not a miracle cell; it's a normal cell that can produce an electrical signal. In the case of plants, almost every cell can do that." A decision-making process occurs constantly within the root tips, but how do they communicate?

The roots act as a network. Do they connect anatomically since it is one root system employing electric signals? Roots emit sounds and electrically generated clicks, but some of those sounds are related to the growth of cells and could be irrelevant. If Mancuso is correct, might plants behave like a murmuration of starlings that paint the sky with undulating liquid forms before they descend for the night, collective avian synchrony based on simple rules and signals? There is no leader, no plan, no direction.

Trees exchange information and nutrients through fungal networks in relatively small areas. Can trees communicate with each other over hundreds of miles? It is a tempting conclusion. Trees reproduce by dispersing their genetic material in ripened seeds such as chestnuts, acorns, redbud pods, maple helicopters, cottonwood fluff, willow catkins, and spiny burrs. Their annual production nourishes white-footed mice, gray squirrels, wild turkeys, red-backed voles, ground beetles, birds, deer, ants, pigs, and crickets. Every few years, a phenomenon known as masting occurs when the same species of trees produce and drop a remarkable crop of seeds simultaneously over a broad area. Masting cycles vary by species and geography. In 2018, oak trees from New Hampshire to Georgia simultaneously dropped millions of tons of acorns, up to ten thousand from a single tree.

What was the signal? How did trees coordinate? There are theories as to the cause of masting. One explanation is that it is precipitated by wind-driven pollination, causing the trees to overproduce. The problem with that explanation is that trees respond upwind where no pollen is received. A compelling hypothesis is that trees coordinate purposefully to overwhelm and satiate ground feeders so that some of the remaining nuts and acorns germinate. But their coordination still needs to be explained. In forests of lowland Borneo, the dominant tree species of dipterocarps respond to El Niño events, which augur drought. In response, forests produce brilliant displays of color where 80 to 90 percent of the trees burst into blossom. A single tree can have up to four million flowers. These events generally occur four years apart, showering the forest floor with ripened nuts and fruits. Forest tribes look forward to this season as they feast on nuts and the nut-fattened wild boar. Although plants respond to the threat of drought by increasing floral production, dipterocarps collaborate simultaneously on over 370 million acres. The trees communicate, evidenced by the El Niño–inspired mass blossoming. They cooperate as a community, a capacity far more developed in the plant world than once assumed.

Is it possible that trees and plants listen in some fashion that we do not understand? Plant hearing may seem unlikely, yet the auditory capacity of plants is not dissimilar to ours. All sound is vibratory. Animals can put their ears to the ground and detect movement and sound. Plants are already in the ground. What do they do with that information? The opposite of hearing is speaking and responding. In the animal world, vocalizing takes many forms. Language is about information, certainty, and survival. Humans talk, chatter, and gossip to share perceptions, learn from others, and reduce uncertainty. For Steven Pinker, language is the "jewel in the crown of cognition." For animals, it is the crown itself. Animals depend on understanding and sharing information about what is happening around them. René Descartes wrote in 1649 that what distinguished humans from beasts was language. It now appears that language and cognitive communication are common to every species.

The belief that language arose de novo from a single genus, *Homo sapiens*, is a form of exceptionalism. Language evolved over hundreds of millions of years before we arrived. The croons, howls, cries, and songs of the Caribbean humpback can be deciphered by whales in Ireland. Elephants rumble in inaudible but meaningful frequencies that can be heard six miles away. Prairie dogs employ adjectives and dialects when warning their coterie of predators. The brown thrasher has a 1,900-song repertoire with vocalizations that sound like flutes and piccolos. Fox yip and gecker with intent and function. Ravens utter a gurgling croak to announce their peaceful presence to other ravens in the neighborhood. Vervets chutter when they see a snake. Mice sing in ultrasound to woo and warn. Research shows that bats distinguish each other by individual names, name their babies, get into arguments, and fight over morsels of fruit. Recent research indicates that elephant rumbles distinguish specific individuals including offspring. The family of life pipes, dooks, wheeks, squeaks, yowls, chuckles, radiates, and caterwauls.

Scientists have largely dismissed the idea of plant intelligence due to the lack of a *botanica lingua*. Is there a language of plants? How could plants be intelligent if the means to transmit knowledge is missing? According to the original definition, *intelligence* means to "choose between," the process of making choices, a capacity infused throughout the living world. Evolution could not have occurred without the binary ability to choose unless you believe life is a random event. The plant community constantly makes choices, altering its behavior moment by moment to survive, if not thrive. Another way to look at this is to ask the question, How did the 438,000 different plants learn to be *that* plant? Look at a meadow or abandoned plot of land. Are the assembled species of vegetation unaware of one another? Unlikely. How do plants interact, link, and learn? The possibility that plants communicate utilizing sound is controversial. Even interspecies communication between animals and insects is discounted, though bears are acutely sensitive to the meaning of grasshopper crepitations and the ascending/descending cry of a redtail hawk. But plants? They don't have a nervous system, brain, or synapses. Although anthropologists and ethnographers have studied Indigenous people who credibly interact with and speak to plants, there has never been any scientific follow-through. The reason is simple: We use ourselves as a reference point for what is possible or not, leading to disbelief and skepticism.

Marine ecologist Monica Gagliano was one of the first scientists to systematically research plant bioacoustics, also known as phytoacoustics, how plants emit and respond to sound. She has written extensively about intraspecies communication and plant cognition. Her interest in the consciousness, memory, and learning processes innate to plants began accidentally in the ocean when she was a reef rat, an affectionate term for people who spend inordinate time underwater at the Great Barrier Reef. She was researching the decision-making processes of Ambon damselfish, a bright yellow species named for the Moluccan island where they were first discovered. The study focused on how mothers pass awareness of a changing environment to their offspring.

Monica visited the same school of fish daily for months. They recognized her and would snuggle into her outstretched hand and allow themselves to be held in her glove. She observed reproductive pairs, offspring, and rituals. Their love songs were recorded, which sound like windshield wipers on dry glass. Besides the wiper crooning, Ambon damselfish use their swimming bladder to produce percussive pops, chirps, and clicks, which describe and guide their behavior. When the research was complete, Monica was required to kill the school, dissect the corpses, remove the organs, and analyze: What was their diet, age, muscle tissue, and reproductive status? She went to the reef on the morning of their last day to say goodbye. There was not a damselfish to be seen. It was a ghost reef, and it stunned her. She knew they knew. The time spent together had broken down a taxonomic boundary. There had been communication, and they had come to know and trust her, but their final message was absence. She returned in the afternoon and euthanized them. Even though the university ethics committee had approved her research, the fish made it known they had not. It was traumatic. She knew she could no longer perform that kind of science. She even questioned whether she wanted to be a scientist.

After completing her paper, she withdrew and spent time in her garden, a retreat that became a turning point. She planted chilies, basil, and fennel, among other vegetables, knowing that basil and chili thrive next to each other. The basil retains soil moisture, acts as a mulch, and emits a protective insecticide for the chilies. Fennel, on the other hand, is allelopathic, emitting substances in the air and soil that inhibit the growth of neighboring plants. Chemical messaging between plants above and below the soil is well documented. When a caterpillar is munching a leaf, the plant synthesizes chemicals to drive the predator away. The defense chemicals release plumes into the air that are responded to by nearby plants that create their own defenses. The compounds vary dramatically, and each molecule has meaning. Fred Provenza writes: "The language of plants is organic chemistry. Each of the estimated 400,000 species of plants on Earth can synthesize hundreds to thousands of primary and secondary compounds . . . from an 'alphabet' of as few as twenty compounds, plants can create trillions of different 'words' by varying the relative amounts of different and secondary compounds."

To determine if plants' signals use other pathways, such as sound, Monica applied the same research protocols to plant studies as she had employed in her highly praised marine studies. She placed potted basil into a separate sealed plexiglass cylinder in the middle of a larger box. A circle of potted chilies surrounded it. The plexiglass boxes prevented contact with airborne chemicals, fungal spores, or roots. There was simply no way a single molecule could be exchanged. The plants were in solitary confinement, encased by a larger outer box connected to a vacuum pump. The sight lines of the plants were blocked by black plastic sheeting. When basil was in the center of the ring, the germination rate of chili seeds was measurably better than when there was no central basil plant. Surprisingly, when fennel was at the center, the chili germinated more quickly, as if the threat of fennel had spurred its growth cycle to accelerate. This is seen in open landscapes. Plants will cooperate, but when threatened by a competitor, they will change their rate and extent of growth. How could the chili plants know whether basil or fennel was their adjoining neighbor? As with damselfish, the question was "How do they know?" In this case, it could only be sound.

It is known that plants emit and perceive ultrasonic acoustic signals when stressed. If dehydrated, a specific sound is emitted. As dehydration persists, plant signals amplify and become ultrasonic "screams" that eventually recede as it nears death. Each plant and type of stress is associated with a specific identifiable sound. Bats, mice, and insects can hear the sounds. The root tips of yellow corn seedlings orient themselves to a recorded sound of water even though actual water is absent. How does the root tip of corn recognize a beneficial sound it has never heard before? Corn roots emit structured, spikelike acoustic chirps. Monica makes a vital point: Her finding did not provide a mechanistic explanation of how the chili seedlings exchanged information with the other plants, but they signposted the way. And what better way to chat with your neighbor behind darkened, opaque, sealed panels of plexiglass than through sound? The ability to sense sound and vibrations is behind the behavioral organization of all living organisms and their relationships with their environment. Why would plants be an exception to vocalization? The obvious objection is that plants do not have ears. However, researchers discovered that plants have tiny hairs on their leaves that are as sensitive to sounds as the hair cells in the inner ear of animals. Monica believes that "humans have something of a track record for silencing those whose voice they do not want to hear. We do this by unconsciously ignoring or deliberately stripping them away . . . the very thing that makes dialogue possible—recognizing the other as an equal." In Monica's case, the botany establishment has done exactly that, even to the point of booing her at conferences. Zoë Schlanger points out that it is only the men who loudly protest and heckle. Yet those same male botanists freely acknowledge that beach evening primroses increase their nectar within three minutes of hearing the sound of a buzzing bee. Science has objectified plants for centuries, so why not a female botanist, too?

Eighty percent of the biomass on the planet is comprised of plants. Using the energy of sunlight, plants transform carbon dioxide and water into glucose, the sugar molecule that runs the world. Glucose is used by the plant to grow more roots, leaves, and stems. As a plant expands, it becomes a glucose factory. The roots provide more water and nutrients; the greater number of leaves produce more glucose. The plant's ability to adapt, generate, and thrive surpasses that of the animal kingdom. Creatures, large and small, including us, depend entirely on glucose made from plants for sustenance and nourishment. Remove glucose from our diet, and every part of our body declines and shrivels. Writes Zoë Schlanger in *The Light Eaters*, "Literally, every thought that has ever passed through your brain was made possible by plants. This is crushingly literal. All the glucose in the world . . . was manufactured out of thin air by a plant."

Plants have been long thought to be inanimate groups of stems and leaves passively transforming light. Botanical science is discovering an extraordinarily different world. Plants have neurotransmitters and a nervous system, yet there is no central organ like a brain to coordinate or register the inputs. As Darwin predicted, they purposefully bend, surround, swerve, climb, and avoid. Plant cells can count. Plants have and regulate their hormones. Plants talk. They send out chemical alarm calls to other plants to prevent predation. Some of the calls are private to their species. Others are universal. Neighboring alders and willows chat using volatile chemicals. Why is theirs not a language? When you touch a leaf or stem, the entire plant is aware and responds (it does not particularly like the familiarity, perhaps to warn itself it may be eaten). You know when your foot itches. Ivy and tomatoes are equally aware, maybe even more so. But how? If touch is repeated and continued, a plant will change its chemistry and structure. How does it make such a decision? What is drawing the conclusion and deciding? What if the plant world is more complex than the human? For fear of censure, some botanists privately ask a radical question: What if the entire plant is a type of brain? What if it knows its surroundings as well or better than an animal and is constantly adapting? Botanists have developed techniques that show how the nervous system of entire plants light up, as do our brains. Might the 80 percent of earthly biomass we call plants be conscious—a stupefying idea to modern science, yet true to traditional ecological wisdom? I remember reading Stefano Mancuso's description of the ability plants have to visually sense motion around them. On the path from my home to the road are twenty-seven redwood trees. Two of the older ones have a nine-foot circumference and extend eighty feet into the air. I had become so accustomed to them that I rarely noticed. On that day, I became aware that they were noticing me. This is the world we inhabit.

Our relationship to the biosphere will determine what lies ahead for humanity. Bending the arc away from blatant degeneration toward ecological recovery depends upon knowledge and respect for the world of plants. Maybe awe is the proper term. They represent the greatest and most exquisite flow of carbon on Earth, the starting point for our breath and life. We might ask, what is more important to the planet, plants or people? If the plants leave, we follow—within days. If we go extinct, plants thrive, and the remnants of our civilization will be covered over by trees, roots, and vines. We like to think we are the most important organism on the planet, a delusion we may want to reconsider.

NINE

Kindom

A spore whose time has come.

PETER MCCOY

Toby Kiers loves fungi. Growing up, she played outside in white smocks and returned home blackened like a coal miner, having pressed her body, nose, and ears to the soil. The varied scents and smells of the ground were seductive, "one of the most complex operas humans are exposed to." As a child, she believed soil held secrets. As a student and researcher, she peered underground into a metropolis of hidden dynamics. She saw plant and fungal resources scurrying in every direction in micron-level interactions at least as complex as what humans do aboveground. They were mycelial networks, the matrix, infinitely connected nodes constantly reshaping the living world above and below.

Fungal communities eat rock, create soil, recycle waste, and alter minds. Everything alive is intermingled with fungi. They are healers of the land, metabolizers and sustainers of life, devouring, decomposing, infusing, and interacting with soil, the atmosphere, and plants. Fungi are the connective tissue of the planet. They had a billion-year head start and are woven into the fabric of all plants, roots, trees, animals, and soil. For more than two centuries, mycology was a neglected science and remains so. Fungi were considered plants and pathogens, but as more was understood about their genes, function, and metabolism, they gained taxonomic recognition in 1969. They were identified as one of the five kingdoms by ecologist R. H. Whittaker. Merlin Sheldrake suggests that fungi might properly be called a "kindom" (kin-dom), the long-overlooked megascience that underpins the other four kingdoms. If anything, fungi are closer to the animal kingdom than plants, but there are stark differences. Fungi digest outside of their body and then absorb the products of their digestion. Imagine putting your body inside your food. Their externalized digestion decomposes the natural world, redistributing energy and nutrients. In a sense, fungi eat death for breakfast to produce life for dinner, the tomb and womb of life. A key message from the kindom is that death is the beginning of life. Without fungi, ecosystems would not exist. And neither would bread, wine, yogurt, beer, cheese, chocolate, and psilocybin.

Toby became a renowned evolutionary biologist researching soil-dwelling mycorrhizal fungi that supply crucial nutrients to around 90 percent of terrestrial plant life. Mycorrhizal fungi exude feathery white mycelia that transform the biology and chemistry of plants and soil in grasslands, forests, and farmlands.

Mycorrhizal describes the relationship between fungi (*myco*) and the plant's root system (*rhizome*).

Mycelia are vegetative threads of fungi, white sticky filaments permeating rotting logs or under wet matted leaves. They are an immeasurable network of continuously branching chains, one cell wall thick, vast underground mats in direct contact with the plant kingdom and the menagerie of organisms that populate the soil. We know fungi primarily as mushrooms, the ephemeral fruiting bodies of belowground fungi that emerge to release spores into the air. Of all species on Earth, their fecundity is unsurpassed. A giant puffball can emit seven trillion spores when it bursts open in a cloudy puff. The parasitic mycelia called artist's conk grows in the heartwood of living and dead trees and can release thirty billion spores a day for months. Fungal fruiting bodies, some delicious and some lethal, are made when the mycelia scrunch together into a thick, cushiony platform. Think of them as fungus blossoms, their way of reproducing. We breathe in fungal spores every time we inhale. Spores can be swept up by winds and storms and travel across oceans. Some spores lapse quickly without food. Viable spores have been found in forty-five-hundred-year-old ice cores. The mycorrhizal fungi that infuse and nourish the soil are arbuscular fungi, which do not release their spores through fruiting bodies or mold.

The marriage of fungi and plants has a storied history. The association began almost half a billion years ago in the Cambrian age. At some point in that epoch, cyanobacteria, better known as pond scum, slipped onto shore, setting the evolution of land-based plants in motion. Fungi had already existed in the ocean for at least seven hundred million years. Under squalls of acid rain, a mash-up between algae and fungi created a pale green mat that spread across the rocky, barren earth. It was the precursor to land plants. Cyanobacteria brought a gift from the sea: it could photosynthesize light and carbon dioxide into energy. Fungi were the perfect partner, able to enzymatically dissolve rocks into nutrients needed for plant growth. Ocean-borne cyanobacteria provide half of the oxygen we breathe today and are the basis for the entire oceanic food chain, starting with delicate zooplankton to baleen blue whales weighing two hundred tons that consume five hundred thousand calories in a single mouthful of zooplankton. As time passed, a growing layer of organic sediment was deposited under rudimentary plant life, a mantle of minerals, carbon, and microbes. The earliest plants, mosses, and liverworts were leafless and rootless for over fifty million years, relying on fungi as their root systems. Horsetails and ferns followed, then trees, forests, and finally grasses, which arrived forty million years ago. Grass seeds—rice, maize, and wheat—supply 60 percent of humanity's caloric intake.

Beneath a typical acre of forestland is a thirty-million-mile mycelial network. That equals approximately a kilometer of strands in a thimbleful of dirt. The nodes and interconnections in the mycelial community exceed the complexity of the internet by a factor of hundreds of trillions. The honey fungus, *Armillaria*, covers four square miles in the Oregon Malheur National Forest and is estimated to be between two thousand

and eight thousand years old. If fungi did not consume what perishes, forests would be littered with a mountainous accumulation of dead logs. The spongy forest floor is due to orchestrated death and rot. Cleaning house is just one of the critical functions fungi perform. Feeding the living world is the other.

The impact of mycorrhizal fungi on soil and plant life is only beginning to be understood. Each plant is a unique quilt of fungi—trillions of fungi-produced compounds innate to plant health. Fungal mycelia form a sheath around root systems. In addition, tiny threads of hyphae penetrate the root tips, providing direct access to water and minerals in exchange for carbon sugars and fats created by plant photosynthesis. Mycelia need energy from sugars and fats; the plants need nutrients, nitrogen, phosphorus, and minerals. Growing tips of hyphae act symbiotically, in sync with the needs of the whole plant. Many desert plants would not exist without these relationships. In an age of increased heat and drying soils, mycorrhizal relationships will become more crucial to plant survival. Current agricultural methods practiced by the food industry destroy mycelia with plows, tillage, herbicides, fungicides, and pesticides. Industrial farming decreases crops' resiliency, nutrient uptake, drought resistance, and nutrient density. Intensively farmed cornfields producing corn syrup, bioethanol, and grain for feedlots are largely devoid of life. The damaged mycorrhizal communities in deteriorated land may require years or decades to be restored. We are going in the wrong direction.

It is estimated that 90 percent of the Earth's soils will be heavily degraded by the middle of the century. Until recently, neither the presence nor support of mycorrhizae was included in the agricultural school curriculum. Yet, fungi determine soil health, which determines plant health and the health of all those that eat plants. The soil's microbiome determines plant nutrients, which feed the human microbiome. Living soil is the primary source of nutrition. Chemical agriculture will never change that; it can only degrade the process.

If you remove a plant from the soil, you will see an asymmetrical reflection of the aboveground plant. When the magnitude of the interconnected fungi is added, the capacity of roots to absorb water and nutrients increases by a factor of ten to one thousand. Additionally, the mycorrhizal networks improve soil structure, reduce nutrient leeching, boost yields, and amplify the availability of phosphorus, nitrogen, and other minerals for the plant. This is not fungal altruism. Supply and demand work in plant/fungal relations; both exchange what they need to flourish.

In her research, Kiers monitored fungal dynamics by attaching nanoparticles to different molecules that emit light when hit with UV light. Once in place, it became visually apparent that a type of trade was taking place. The trading strategies between plants and fungi were complex and sophisticated. Decisions and tactics were taking place at extraordinary levels of complexity. A single plant can have a million root tips connected to mycelial hyphae, operating independently but not autonomously. According to Kiers, fungal transactions appear similar to marketplace transactions. The fungi seemed to have a "buy low, sell high" strategy where they would stock up on phosphorus when it was abundant in one area and exchange it to plants where it was scarce in exchange for richer carbon rewards. Toby conducted an experiment where roots deficient in phosphorus were placed in an area not far from well-nourished roots. When the mycelium detected the lack of phosphorus in one group of roots, it moved phosphorus to those areas, receiving a higher carbon payment. If plants had abundant phosphorus, fungi accepted relatively low levels of carbon as payment. The exchange rate varies and is calculated by both plant and fungus, not dissimilar to the end of the day at a farmer's market when vendors discount produce prices so as not to return home with crates of tomatoes.

Kiers founded the Society for the Protection of Underground Networks (SPUN) to map the biodiversity of the Earth's mycorrhizal communities. SPUN is activating mycologists around the world, working with local collaborators to obtain core soil samples from all types of ecosystems. The goal is to survey, locate, and identify the diverse domains of mycelia and create a fungal atlas, a biological reference guide to ensure their protection and continuity. You cannot move fungi into different soils and ecosystems, as can be done with plants and animals. They are unique to the place they are found and change as their environment changes. Chilean mycologist Giuliana Furci's research shows that fungi on the bark of a 400-year-old tree are not the same as those on a 390-year-old tree. According to mycologists at the Royal Botanic Gardens at Kew, less than 10 percent of fungi species are known to science, leaving 2.2 to 3.8 million more species to be identified, approximately 30 percent of species on Earth. Kiers is mapping what has been called the "dark matter" of life on the planet.

A recent peer-reviewed study by Kiers and colleagues estimates that plants direct the equivalent of 13.2 billion tons of carbon dioxide annually into mycorrhizal mycelia. This is a conservative estimate, not far from the combined annual carbon emissions of China and the United States, the world's two largest emitters. This is a key entryway for carbon, much of which is taken up by organisms in the soil. Fungal carbon capture is almost entirely ignored while mechanical methods of direct air capture (DAC) of carbon make for breathless headlines. The Stratos carbon capture plant being built by Occidental Petroleum in Ector County, Texas, will be the largest of its kind thus far, a forerunner of one hundred more such plants on the drawing board that "suck" carbon dioxide from the air, liquefy it, and pump it into underground geological formations. Its estimated cost is $1 billion. It will require $300 million annually to operate and will draw down five hundred thousand metric tons of carbon dioxide annually, what fungi sequester in nineteen minutes, what the world emits in 293 seconds. The plant, known as an atmospheric carbon scrubber, does what biological systems have done for billions of years. If every US household with an electric clothes drier air-dried their wash six or seven times a year, it would reduce more greenhouse emissions than the Ector County plant and save households $3 billion in electricity bills.

Complex carbon flows are seen racing in both directions within mycelial networks when observed under Toby's microscope. How is the flow of carbon determined and regulated? No one knows. The carbon being fixed is complex, with longer chains that can remain in the soil for hundreds, even thousands of years. Annually, fifty-four billion tons of greenhouse gases migrate into the atmosphere due to human activity. At the same time, mycorrhizal fungi build long-term carbon reservoirs beneath the soil. The estimated two and one half billion tons of carbon in the Earth's mantle is over three times greater than carbon in the atmosphere. At the same time, industrial agriculture destroys fungi on hundreds of millions of acres of farmland. If we continue to over-till, overgraze, and place herbicides and fungicides on productive land, mycelial networks will further recede, forfeiting their capacity to lock carbon into the soil. If we lose 8 percent of the 2,500 Gt of terrestrial carbon, atmospheric gases, already at a 4.3 million-year high, will increase by 100 ppm.

Walking on the ground, we traverse an active, vibrant, intensely energetic kingdom of life. There is more life underneath than above the ground. Although there are thousands of studies on mycorrhizae, and their diversity, behavior, and locations, mycorrhizal fungal networks remain more enigma than fact. Most research is done in controlled settings, under microscopes, or in pots. How does one research if mycelia have one million distinct touch points with a single plant? Or how do you study the Malheur *Armillaria* fungal parasite that covers ten square miles and weighs seventy million pounds?

Mycorrhizal networks act as if they can solve complex calculations. Studies portray mycorrhizal networks in terms of exchanges, symbiosis, chemistry, and processes that can be measured, observed, and understood. The fungal world underfoot is bewildering, labyrinthine, aware, and sensate, with untold trillions of intersecting connections transferring molecules and information. Yet, fungal networks have no brain or nervous system. One theory is that they *are* loosely analogous to a nervous system and utilize amino acids, as do our bodies, to transmit information. The most crucial way fungi communicate is sexually, finding a mate in order to reproduce. They use the equivalent of their nose and either wend their way toward or away from potential mates, depending on their pheromones. Sounds familiar.

Voyria, commonly known as ghostplants, do not photosynthesize and still receive plant carbon sugars via mycorrhizal networks. Ghostplants earn their name by seemingly returning nothing. There is no obvious symbiosis or reciprocity. They break the rules when it comes to mycorrhizal networks. Plants exchange the carbon they receive, but why would plants or mycorrhizae share and not get anything in return? Until recently, plants were considered standalone, competing entities, studied and analyzed accordingly. From an evolutionary perspective, nourishing *Voyria* would be termed altruistic, but altruism eventually fails because it rewards the takers and punishes the generous. Mycorrhizal networks and billions of plant roots are a community. The question is, What are the dynamics of those communities?

When Kiers studies the complex communities of mycorrhizal networks, they look more like cities with rush-hour traffic, nutrients and chemicals moving in multiple directions inside microtubule networks. Nutrient flows might go one way and then reverse, stop, continue their flow, slow down, and then reverse again. It is directed and rapid, a constant, almost herky-jerky response. But a reaction to what? There is a clue. Mycelia emit both chemical and bioelectrical pulses. Kiers works with Cosmo Sheldrake to record sounds shunted across underground ecosystems in mycorrhizal biodiversity hotspots such as Chile. They are mapping out how fungi sense, learn, and make decisions. Fungi are polyglots, speaking and understanding a wide variety of chemical signals. In Merlin Sheldrake's superb treatise on fungi entitled *Entangled Life*, he points out how the human obsession with animals created what is known as "plant blindness, where trees are seen as scenery, a backdrop to deer and birds." Sheldrake asks whether we also have "fungal blindness."

A deluge of studies reexamines prior beliefs about consciousness and intelligence in animals, birds, plants, and insects. Previously unthinkable conclusions are commonplace. Insects have cognitive capabilities once reserved for higher primates. Bees count with numbers, remember human faces, have dialects, and only see green when flying but full color when they land on blossoms. Wasps can recognize individual wasp faces. Sleeping jumping spiders might be dreaming when they hang upside down on a thread, curl their legs, and twitch. Clark's nutcrackers remember where they buried thousands of pine nuts in up to ten thousand caches within a fifteen-mile range. Crickets teach offspring about the danger of spiders before birth, with their mothers long gone. Ravens can pass on the memory of a human face to their progeny. Farmer ants play, grow food, and herd aphids as livestock. Do we know what fungi are doing?

The above descriptions are accurate but anthropomorphic; they describe intelligence within our framework of understanding. As brilliant as *Homo sapiens* are, how do we assess the intelligence of other species? There are principles and constants in the inanimate world of physics and chemistry, but biology defies principles. Life sciences have been incorrect for five hundred years, and it is supposed to be that way. Science records what is accepted and known at a given time, not a frozen truth. Samuel Wilberforce ridiculed Thomas Huxley in the famed 1860 Oxford debate on evolution by asking Huxley which side of his family tree descended from apes. By its nature, science sticks to paradigms until proven otherwise. In the seventeenth century, animals were seen as machines: no feelings, no emotions, no thoughts, and they were certainly not relatives.

If species perceive the world in a way we cannot, how would that influence their logic, memories, and communication? When a blind bat echolocates using ultrasonic sound, a dragonfly sees the world in wraparound panoramas, or a bee detects ultraviolet footprints to know when a blossom was last visited, they occupy a distinctly different world. It is difficult to fathom mycorrhizal fungi as intelligent. Plants are capable of cognition, communication, computation, learning, and memory, yet have no brain, neurons, or nervous system. Might fungi be similar? Some say that without consciousness, there can be no intelligence. Since science does not know what consciousness is, there may be better starting points. Can we know what other species know, and by inference, can we understand them or ever understand their language? As Toby Kiers says, "It is crazy what's going on down there."

TEN

Parlance

*Earth languages are not lies or manipulations to
serve political, religious, economic, or scientific ra-
tionalizations. They are not invoked, entrusted,
or gifted to be placed within linear boxes of data.
They are spoken every moment as cures, where
all praise goes to the Earth.*

TIOKASIN GHOSTHORSE

It is commonly believed that we talk about *what* we see. However, what we say about ourselves and the world represents
how we see. In times of political cacophony, this is obvious. We
no longer expect the truth. When we talk about experiences,
our thoughts enter two prisms. The first is the self—our iden-
tity, purpose, and beliefs. The second is our learned language.
Our native tongue determines how we see the world. In Japan, an
envelope is addressed the opposite way as done in the West. It
starts with the country and ends with the first name. The sense
of being a small part of a larger system is embedded in the lan-

guage. The individual is subordinated to the whole—respect and humility are stitched into the language.

English is an explicit language that originates from the self. Sentences often start with the first person, whereas Japanese sentences usually have no pronoun. Any culture can speak English, but the English language does not derive from one culture. It is a polyglot language made of words from other languages, people, and times: intertwined is Greek, French, Latin, Anglo-Saxon, Teutonic, Danish, Dutch, German, Celtic, and Italian. That is why the pronunciation rules of English make no sense. *Steak* does not rhyme with *streak*, *some* with *home*, or *head* with *heat*. An estimated 160 English dialects currently in use constantly create new English words. *Ta, brekky, arvo, strewth,* and *okker* come from the "strine" slang of Australia. Gullah English, spoken in coastal South Carolina and Georgia, combines words from Bantu and the Niger-Congo. *Chany* is chinaware, *mannusstubble* is polite, *crackuhday* is dawn, *ceebin'* is deceiving. The dialect introduced *dunno, no-count, granny,* and *gimme* into common parlance.

English is the language of commerce, dominant in science, ubiquitous in IT, and preeminent in technology. All of the world's one hundred million synthesized chemical compounds are registered in English, making industrial chemistry the most prominent language group (by word count) globally. The second-largest language by word count is plant names. Then comes English, Mandarin, Hindi, and Spanish. English is the connective medium for a vast number of international conferences. Thousands of new words are added to the English language annually, making it even more precise and useful. It has named, identified, and labeled virtually all known things alive and lifeless, a stunning achievement. However, the categorizations contained in the one million words of the English language leave speakers without a framework, context, or sense of place. It is rootless. Language can either connect or disconnect one from life. Notice that when we see a deer or owl, we say *it* nibbles the grass. Or *it* is wheeling above. We would never do that to a person. It eats? It drives? It graduated?

There are other ways to speak and see. Language is about information, connection, certainty, and survival. The Chicham language of the Achuar in the Amazonian region of Ecuador does not have a word for nature. Nor do other Indigenous languages. There are good reasons for the absence of concepts like nature. Such words would only be needed if the Achuar experienced nature as distinct from the self. Harvard researcher Andrew Messing commented that looking for the word *nature* in Indigenous languages is like looking for the word *mansion* among forest dwellers.

I once witnessed a science panel on natural resource management practices and noticed that the one Indigenous panelist, Oren Lyons, a Faithkeeper of the Seneca Turtle Clan, was quiet. In the spirit of inclusion, when the moderator finally turned to Lyons for his thoughts, he said, "In our culture, resources are relatives." The six words were said without judgment. It was a matter of fact. Lyons's perspective was in contrast to what had preceded it, an intimate way of being within the living world as contrasted to seeing it as an object.

When journeying to Torres del Paine, Patagonia, I happened to walk into the unheated Yámana Museum in Ushuaia, Tierra del Fuego, on a wintry day. It was hardly a museum. It was cold and dank inside. No one was there. I wandered around the two small rooms and stared at black-and-white photographs that seemed to stare back. The images were mostly of semi-naked people, often protected by seal fat and loincloths, gazing apprehensively at the camera. The nineteenth-century photographs showed Yámana families huddled in habitats made of sticks, grasses, and sealskin. The tribespeople had high cheekbones, strong jaws, short, powerful bodies, thick mats of raven hair, and ebony eyes that spoke of fear and distress. An invading settler culture had destroyed their way of life and was now photographing them as objects. The museum was a remnant of their existence. It was poignant, painful, and heartbreaking.

In 1520, three caravels captained by Ferdinand Magellan rounded a prominent headland at the fifty-second parallel in southeastern Chile. He had discovered the rumored southwestern passage to the Pacific Ocean, a strait that now bears his name. The fleet sailed into an unforgiving land. Copper-colored people watched Magellan's ships on bluffs amid the dense beech forest. It was October, still wintry in the southern hemisphere. The rapt audience, slathered in seal blubber and occasional animal skin, was indifferent to the gales that chilled Magellan's sailors to the bone. Standing upon the windblown ridges were people who had dwelled at the "end of the earth" since the last ice age. They lived in pit houses, wigwams, and tree-bark canoes and maintained no permanent homes. To ensure their survival, the Yámana carried baskets of embers wherever they went, a constant lifeline to fire. At sea, they placed coals on sand in the canoe's center, a welcome heat source for the women who would dive into the forty-eight-degree water for shellfish. The Yámana could sleep outside at night naked and unsheltered. Sailors shivering on the ships watched sleet melt on the skin of topless men and women. The smoky fires Magellan saw across the archipelago gave rise to the name Tierra del Fuego.

The last speaker of the Yámana language, Cristina Calderón, passed in 2022, at age ninety-three. The impact of enslavement, disease, racism, and exploitation resulted in the extermination of the Yámana. Yet, in this case, there remains an unusual legacy. For over twenty-one years, their language had been meticulously cataloged and recorded by Thomas Bridges, who tended to a small remnant population of the Yámana at his mission. As an amateur lexicographer, he worked with the last tribal head, George Okkoko, documenting the meaning of 32,240 words before Bridges himself passed in 1898. We do not know what the final total would have been had he been able to complete the dictionary. There are far more words in the Yámana lexicon than Shakespeare utilized in his career. Educated Americans have a vocabulary of approximately 20,000 words. Teenagers use about 1,200 words.

I found a copy of the Yámana-English dictionary (three hundred were printed in the 1950s by Austrian anthropologist Martin Gusinde). Reading it is like entering a different realm of existence. It is a guide to the hauntingly beautiful land where the Yámana thrived and Westerners survived. It conveys the richness of their language and daily life. In Yámana, *Taisasia* means to be covered up and lying on the ground like eggs in a nest. The word for depression is a crab that has not fully molted its shell. *Ondagumakona* means plucking clusters of mussels from a rock on a boat and cooking and eating them simultaneously. The Yámana language was a teaching—it explained how to be a successful resident on the land and seas they lived upon. It taught the Yámana where they lived. The word *Yámana* means the highest form of life, to be alive.

One in five people in the world speaks English. If Mandarin, Hindi, and Spanish are added, that comes to half of humanity. Another 7,145 languages are spoken; about 40 percent are considered endangered, with less than one thousand speakers. According to UNESCO, 608 are critically endangered, with less than one hundred speakers. There are 154 endangered native languages in the United States, including Assiniboine, Chickasaw, Seneca, Salish, Tewa, Tlingit, and Yurok. A language dies when it is not spoken to children. What value could these languages have today in a rapidly modernizing, globalized, educated world? The truest answer is that we do not know because there is no "we." Who is to judge and decide what is or is not of value? The languages that are perishing are specific to people, cultures, and places. Hundreds of languages have followed displaced cultures into urban environments, usually by people seeking work. In New York City, more than 700 languages are spoken, of which 150 are endangered, the greatest concentration of threatened languages in the world. Several hundred speakers of Seke reside in and around two villages in Nepal. Another 150 Seke speakers live in two apartment buildings in Brooklyn. There are more languages per square mile in Queens than anyplace in the world. Populations of speakers in New York City include Bartangi, Mojave, Taino (the people nearly extirpated by Columbus on his first voyage), Chamorro, and Sorani Kurdish.

Anthropologist Wade Davis believes the languages spoken worldwide represent an *ethnosphere*, "the sum total of all the thoughts, dreams, ideals, myths, intuitions, and inspirations brought into being by the imagination since the dawn of consciousness." In addition, they represent, as in the case of the Yámana, an extraordinary record of observational science, the accumulated wisdom of living in and learning from one region, island, or forest for thousands of years. Languages consist of more than verbs, nouns, and adjectives. They contain teachings, mores, and rules developed by trial and error over eons. Each language is a way of seeing the world. The state of the world reflects how dominant languages have overwhelmed the extraordinary diversity of human culture. Aboriginal people of Australia refer to themselves according to their home, such as Koori, Gandangara, or Nunga. Author Clare G. Coleman describes what Aboriginals do when they meet for the first time: they discuss family going back twenty generations or more to determine how they are associated and linked. From their point of view, what makes us human is how we are connected, not what we have. Davis considers languages in the way a biologist sees species diversity. "Distinct cultures represent unique visions of life, morally inspired and inherently right. And those different voices become part of the overall repertoire of humanity for coping with challenges confronting us in the future. As we drift toward a blandly amorphous, generic world, as cultures disappear and life becomes more uniform, we as a people and a species, and Earth itself, will be deeply impoverished." There is a language ecosystem in the world, and it is no coincidence that the density and location of Indigenous cultures closely match the remaining areas of greatest biodiversity.

The Mi'kmaq live in Canada's northeastern woodlands down to the northern tip of Maine. They name large pine trees by the sound of the wind moving through the branches one hour before sunset in October. Years later, elders can remember the names given to native stands of pine and detect if trees have been damaged by comparing their names with current sounds. We have several words that describe the sound in trees—soughing, rushing, sighing, and moaning. But English, with its one million words, has no name for the sound made by an individual tree. To name and remember a tree by its sound is an extraordinary linguistic feat. It is a proper name based on a unique sound and is untranslatable into any other language.

On a flight to Alaska, I was seated next to a Yup'ik woman who was going home because her sister had passed, and she was now the family elder. The Yup'ik live along the Bering Sea in western Alaska and Bristol Bay. Their ancestors arrived ten thousand years ago from Siberia, and their direct descendants have lived in western Alaska for the past three thousand years.

I asked what it was like to live on the Bering Strait year-round. As she recounted stories of a seminomadic life, she mentioned in a matter-of-fact way that they could predict the weather two years in advance, an ability that was crucial to their survival. It helped estimate how much dried fish and seal meat to put away, how long the winter months would be when they could move inland, and what the caribou population would be. She continued to share details about hunting, winter attire, and birdlife, but I had to interrupt her there. I pointed out that over one hundred geostationary weather satellites cannot reliably predict weather within a week. Two years in advance?

She explained that understanding long-term weather cycles was based on daily observational awareness for the Yup'ik. Specific events and occurrences were remembered and passed down through generations about when and what occurred in their environment and what would follow in the months and years after. This included when the sea ice froze in the fall or thawed in the spring, the color of the ice when it froze, its texture and strength, the texture of velvet on the caribou antlers, the color of the tundra moss, the arrival of bowhead and beluga whales, Pacific walrus, and bearded seals, types of snowfall, the texture of convective and stratiform clouds, and the timing of auklet and eider migrations. Over thousands of years, observations of life and the elements were correlated to later events. This acute awareness of place is pattern recognition, the connection between memory and current information, or, in this case, weather. Yup'ik does not have more words for snow, moss, or clouds than science. The difference is how they use words to create an enduring culture. Scottish has hundreds of words for snow. *Flindrikin* is a light snow shower, *blett* are large snowflakes, *smirr* is sleet, and there are four hundred more words. The language is colorful, charming, and localized. But it was never synthesized into patterns that foretold the weather years hence.

Between November 11, 1861, and January 24, 1862, California experienced historic rainfall that became a megaflood, a deluge that continued for forty-three straight days. An unrelenting atmospheric river from warm Pacific waters left devastation in its wake. Thousands of lives were lost. Homes, settlements, ranches, and livestock were swept away. The Central Valley became an inland sea that submerged towns under six to fifteen feet of water for six months. One third of the property in California was destroyed, and the state was forced to declare bankruptcy. The governor had to row into the second story of the state capitol in Sacramento. People were caught by surprise, except the Maidu tribe. The *Nevada City Democrat* recorded their response on January 11, 1862. The Maidu left their homes "for the foothills, predicting an unprecedented overflow. They told the whites that the water would be higher than it had been for thirty years and pointed high up on the trees and houses where it would come. . . . It is not improbable that they may have better means than the whites of anticipating a great storm." They did—ancestral memory developed over two thousand years of habitation in the Sierra foothills and inland valleys. At that time, five hundred thousand people lived in California. Today, that number tops thirty-nine million.

Geologic evidence shows that the 1861–1862 megaflood has occurred periodically, every one hundred to two hundred years, for thousands of years, and will likely occur again.

Though English has an extraordinary number of words, that does not mean words from other languages can be translated into English. It may be a universal language, but despite its breadth, it forsakes being a language that teaches where one lives. In that sense, it will always be homeless. Indigenous language befits place and people, a means to ensure the community is connected to each other and the region. Wade Davis's concerns are accurate. Humanity does not realize that it is losing and eliminating Indigenous languages, how the languages reflect a unique and brilliant understanding of the world. As impacts from a warming atmosphere pummel neighborhoods, rivers, forests, people, and towns, we will need to know our homelands far better than we do now—the biome, where our water and food come from, how to create durability, foster antifragility, and deepen connections.

The climate movement employs words and phrases that make little or no sense to the vast majority of humanity—*net zero*, *decarbonization*, *direct air capture*, *enteric fermentation*, *carbon removal*, *teragram*, *tipping point*, *planetary boundaries*, *sequestration*. The most outlandish term of all may be *carbon neutrality*, a biophysical impossibility. This has been called "nounism," a particular way of dissociating from the world, where divisibility becomes analogous to knowledge. What people respond to is verbs. Lakota teacher Tiokasin Ghosthorse explains why Indigenous languages like Yámana have more verbs than English. They are about relationships. In contrast, nouns divide the world into things. The collapse of living systems has roots in grammar and vocabulary. Although Indigenous people have been forced off their tribal lands for centuries, Ghosthorse points out that the land could not be taken away from them because they never believed it was theirs. Tribes and cultures can be landless but not homeless. They did not live on the noun we call *land*. William Least Heat-Moon recounts an old Indian story:

> The white man asked, *Where is your nation?* The red man said, *My nation is the grass and rocks and the four-leggeds and the six-leggeds and the belly wrigglers and swimmers and the winds and all things that grow and don't grow.* The white man asked, *How big is it?* The other said, *My nation is where I am and my people where they are and the grandfathers and their grandfathers and all the grandmothers and all the stories told, and it is all the songs, and it is our dancing.* The white man asked, *But how many people are there?* The red man said, *That I do not know.*

Most of the world believes land can be taken and owned, and the results show. Half of the world's farmlands are degraded. One third of forestlands have been eliminated since 1800, along with 60 percent of grasslands. Swaths of farmland and water rights are being scooped up by multinational corporations and foreign countries. Forty million miles of roads cover the land, enabling US trucks and automobiles to kill more than one million vertebrates daily. Ghosthorse suggests that the modern era may have exceeded its "sell by" date. Popular modern discourse is a mental goo of beliefs and hearsay. "Conscious languages do not require a *logic of believing*, but rather a *logic of knowing* that the Earth does not lie and only speaks the truth with conscious respect for all Beings."

ELEVEN

Paper Eyes

In the name of the Bee,
And of the Butterfly,
And of the Breeze—Amen

EMILY DICKINSON

A flame skimmer hovers two feet from my face, looking straight at me. Its red, bulbous eyes have twenty-four thousand corneas, allowing it to see 360 degrees—up, down, backward, forward, and around simultaneously. What it makes of me visually is unimaginable. It has thirty opsins, the universal photoreceptor molecule that resides in the visual systems of the animal kingdom. I look back with my three opsins and two blue corneas. At the fishpond where I am sitting, the skimmer hovers and dashes about with its satin wings, red and orange, body and wing. My visitor weighs a tenth of an ounce and darts around at speeds up to 30 mph for the few weeks it lives in the air. Its three- to four-year lifespan is mainly spent as larva, a freshwater nymph, an underwater omnivore feeding on tadpoles, spot tails, minnows, and other nymphs. Today, it is sporting iridescent wings that sparkle like ball gowns, looking to mate, which it prefers to do immodestly in the air. As it sparkles and pulsates in front of me, perfectly stationary, I am looking at 350 million years of evolution.

Dragonflies' compound eyes see ultraviolet light, giving them a unique ability to detect shape and movement. Military experts have studied their behavior to develop software for stealth aircraft because they employ active motion camouflage. When a dragonfly hunts, it hovers perfectly still and positions itself between its prey and a shadow cast behind it by, say, a tree, concealing its position. It's a bit like creeping up on someone in a forest by hiding behind branches. The dragonfly can continually change its position based on where its prey moves to keep the tree and its prey lined up—a butterfly or mosquito won't notice. The dragonfly gradually looms closer until it reaches striking distance. This evolutionary adaptation has suited dragonflies well, and they have become highly successful predators despite living short lives.

A remarkable example of Indigenous observational science was revealed in 1949. Ethnoentomologists documented how the Diné (Navajo) named and classified over seven hundred species of insects, describing their sounds, behavior, and habitats, knowledge that had been shared, memorized, and passed on for generations. Why did the Navajo do this? Maybe because they *are* scientists. They wanted to know their world better, knowledge that could be the difference between surviving and thriving for people who live completely on and with the land.

In the forests of Mexico resides the giant owl butterfly (*Caligo eurilochus*) that flaps its papery seven-inch wings at dusk when predator birds are absent. At the base of each wing is a perfectly formed eyespot that, together, uncannily resemble the eyes of an owl. An artist would be challenged to create such replicas. English naturalist Henry Walter Bates first explained the copycat wings. Arriving in 1848, Bates traveled far up the Amazon and its tributaries. Aside from its racist tropes, his book, *The Naturalist on the River Amazons*, is a stunning description of eleven years of study wherein he collected over fourteen thousand species and identified eight thousand new ones, from mundane foraging ants to jaw-dropping, foot-wide, bird-eating *Mygales* spiders that giggling children paraded around leashed like a pet dog.

Like other naturalists of his day, John James Audubon and Alphonse Dubois, Bates shot birds right and left to be packed in formaldehyde and handed over to natural history museums. However, his primary interest was Amazonian butterflies. He noticed edible butterflies being ignored by insectivorous birds and dragonflies because their wing coloring mimicked noxious or predatory species. The spicebush swallowtail caterpillar is born black and white, disguised as bird poop. It will molt three more times with spots on its head that make it appear like a snake—known as Batesian mimicry. This evolutionary process rewarded deceptive wing patterns and coloring to protect species.

He was an early supporter of Darwin's theories of evolution, and Darwin called Bates's book the best on natural history ever published in England. What neither Darwin nor Bates could explain is how butterflies manage to disguise themselves. Had caterpillars ever looked at an owl's eye? They evolved, to be sure, and failures are gobbled up, but how exactly does the pupa of a caterpillar morph into a butterfly with perfect replicas of owl eyes on its wings? The scientific explanation is a regulatory network that allows genes to collaborate and learn from each other. That does not tell us how the genes were programmed in the first place. Millions of years ago, genes began to paint wings with pigmented designs of extraordinary fidelity and complexity. Who was the artist?

When it comes to butterflies, most people are aware of relatively few species. Particular attention is paid to monarch butterflies that migrate two thousand to three thousand miles from their summer feeding grounds in northern America and southern Canada to the oyamel fir trees in central Mexico, where they overwinter. The brilliant black and orange wing patterns that make them a spectator favorite are intended to remind birds that they are bad news. Monarchs lay their eggs under the leaves of the toxic milkweed plant. When caterpillars hatch, they consume the milkweed leaves, so the monarch emerging from the chrysalis is equally toxic.

Lepidopterists who study butterflies and moths have a panoramic view of their importance and presence. While there are 19,500 species of butterflies, there are more than 160,000 species of moths, many of which are similar to butterflies in shape and wingspan. However, they vary widely in size, from pygmy moths with the wingspan of a spaghetti strand to Hercules moths that are a foot wide. Moths and butterflies are of the same order, even to the point where some scientists call butterflies daytime moths. Recent studies show that moths can be more effective pollinators than bees. They do this mostly at night. They blanket the Earth when we aren't looking. To explore their ubiquity, you need to start at dusk, ideally on a moonlit night in the absence of any artificial light. Let your eyes adjust, wait, and watch. It is revelatory. I went out one moonlit night to pick a sprig of rosemary, and there were dozens of slender, silvery visitors on every branch. Moths have extraordinary antennae that can detect the scent of flowers or prospective mates miles away. Whereas bees are seen as the dominant diurnal pollinators, moths touch upon a greater variety of blossoms at night. Moths' herky-jerky flight patterns are defensive maneuvers designed to outwit the echolocation of predator bats.

One in ten described organisms on Earth is a moth, and more than 90 percent of birds feed upon them. On the other hand, many moth caterpillars depend on specific families or species of plants. Due to habitat loss, plants are disappearing, removing a species of moths forever. A circle of compounding loss occurs as moths, plants, and birds disappear. The next time you see a bottle of mezcal, note that the preserved worm at the bottom of the bottle was a wannabe moth.

The diversity and adaptations within the insect world may be more profound than we have come to believe. Could insects be sentient, aware, conscious of us, and capable of feeling? According to recent studies, they can. It is assumed that tiny brains do not support sentience, but that belief does not hold up. The same midbrain and basal functions that support our awareness of the world are represented in invertebrates. There is a "minds without spines" movement that questions why morality and animal welfare stop at invertebrates. Concerning bees, much is coming to light. According to zoologist Lars Chittka of the University of London, studies show that honey bees (*Apis mellifera*) can count, make contrasting distinctions, and learn by observing others. They experience pain and pleasure, are aware and conscious of their knowledge, and recall past stories, as do we. That may seem unimaginable in a brain that weighs two to three milligrams, a tiny fraction of the million-milligram human brain. However, each nerve cell in the bee brain can make connections with ten thousand other cells, providing more than a billion connection points in the bee brain at any given moment.

Inside active hives is a din of buzzing. When researchers injected small probing microphones within hives, the cacophony turned out to be short bursts of encoded information that alert bees to nectar location, food quality, and distance above the ground. Chittka, who has devoted his life to studying bees, believes that their sense organs perceive the world in such profoundly different ways from what we can grasp that they "might be accurately regarded as aliens from inner space." They have near wraparound vision with their opposing bulbous eyes. Their entire diet is contained within a flower; the range of color spectra they can see far surpasses our own; they have a magnetic compass in their tiny brain and protrusions on their head, which can extend two feet, that taste, smell, hear, and sense electric fields. And they are precision pilots. Chittka asks the unanswerable: What is on their minds? "It now looks like at least some species of insects, and maybe all of them, are sentient."

Nearly 1.1 million species of insects have been identified. Might others be as alien and aware as bees? Why not beetles, cave spiders, locusts, and skimmers? For each human being on the planet, there are approximately 1.4 billion insects weighing in at one thousand pounds per person. Insect populations are down 30–75 percent in the past four decades, perhaps the most significant disappearance of life since wooly mammoths vanished ten thousand years ago. Globally, there has been an annual 2 percent reduction in insect biomass over forty years with no end in sight. This is responsible for the loss of three billion birds. Without the creatures that buzz, fly, swoop, hum, burrow, bite, and munch, most birds cannot exist. Neither can most plants. Eighty percent of wild plants depend on pollinators. Insects go away, we go away. The food chain would be irreparably broken. If there were few or no pollinators, animals, birds, and fish would cease to exist within months. Agriculture would follow within a year as there would be few functioning farms. Oceans would take a couple of years longer. Fungi would hang on for a few years to clean up the decomposing corpses, then they would disappear. The Earth would regress one billion years and become a near-barren planet of bacteria and protozoa. The counterpart to E. O. Wilson's famed 1987 paper, "The Little Things That Run the World," might be "Without the Little Things, the World Does Not Run." According to Dave Goulson, professor of biology at the University of Sussex, land is becoming inhospitable to life.

Insects are integral to terrestrial ecosystems, performing irreplaceable ecological services. They provide natural checks on some of the destructive insects and assist in the decomposition of leaves and wood, soil formation, water purification, and carbon sequestration. Insects break down and consume natural waste, including feces, biomass, and corpses. They pollinate 1,200 crops and 180,000 plant species. Insects feed fish, birds, pangolins, reptiles, and bats. And yes, they are pests, carry disease, devour crops, and drive you nuts at night, buzzing around your head to suck blood for a tasty meal. Understandable aversions notwithstanding, protecting insect habitat is crucial to the survival of tens of billions of birds, reptiles, fish, and mammals and is indispensable to ecosystems.

The relative climatic stability we still enjoy is thanks to woodland, slough, prairie, bog, meadow, delta, grassland, cripple, taiga, coral reef, mangrove, salt marsh, and tundra. These systems draw down and store billions of tons of carbon from the atmosphere annually. Insects depend upon them, and vice versa. Ecosystems are buffers, biological reserves holding three billion tons of carbon above and below ground, four times more carbon than is in the atmosphere. Without beetles, butterflies, and bugs, ecosystems stagnate, shrivel, fade, wither, turn to mush, and perish. This potentially catastrophic dynamic is easily ignored because it cannot be seen. Scientists who study earth systems believe the insect crisis poses as severe a threat to humanity as the climate crisis.

Insect collapse was notably detected by scientists working as amateur entomologists on weekends. Scientists make discoveries that can surprise a layperson. In this case, laypersons shocked the scientists. A study that rattled the world's entomological community was done by the amateur Krefeld Entomological Society in Germany. It had kept meticulous records beginning in 1905, collecting nearly one million insects from the same nature reserves in North Rhine–Westphalia. In the 2000s, they identified startling declines in biomass in the traps used to gather flying insects. From 1989 until 2016, the measurable biomass of flying insects declined by 76 percent. Their data, first published in 2013, was later dubbed the "insect apocalypse" by the media. The news spread and scientists around the world confirmed their findings.

The downturn didn't need science to be revealed. Farming regions around the world are witnessing insect collapse. When I was young, streetlamps were mobbed and encircled by moths—black witches, loopers, owlets, skippers, and common buckeyes—but no longer. I grew up in the agriculturally rich San Joaquin Valley; my uncle would stop his car after a few hours of night driving, take out a metal ice scraper, and remove the splattered insect protein from the windshield. It was a menagerie plastered against the grill—grasshoppers, glasswings, damselflies, bumblebees, stoneflies, swallowtails, and hooktips. Insects were so profuse that wire mesh was placed on car grills so that radiators did not overheat. Decades later, I travel the same highways with a clear windshield. The "windshield effect" is seen the world over. Birds are declining because 96 percent of their food is insects. The future of insects, birds, and humanity rests in the hands of the agricultural food system because it is the greatest emitter of poisonous chemicals into land, air, and water. The preponderance of research on insect life is focused on how to kill them.

Pesticides and food abundance were promoted as a simple calculus. Agrochemicals promised higher yields for the farmer when relevant toxins were used. Farmlands became repositories for seventeen thousand types of pesticides, herbicides, and fungicides. The herbicides paraquat, dicamba, and glyphosate destroy unwanted grasses and weeds. Methyl bromide, organophosphates, and chloropicrin fumigate the soil. A jar of US applesauce for children will likely contain acetamiprid, fenpropathrin, carbendazim, and some sixteen other pesticides. The most damaging insecticide used on farms today are neonicotinoids, a class of pesticide invented to replace the morbid toxicity of other pesticides. It binds to insects' nerve cells, causing paralysis and death.

Acetamiprid, the dominant chemical in applesauce, is a neonicotinoid. When used as seed coatings, 5 percent goes to the crop. The remaining 95 percent goes into the soil, roots, grasses, streams, and rivers, where it remains for five to six years. Insects perish when they nibble or pollinate plants that have taken up neonicotinoids from soil or water. Nearly 75 percent of flowering plants depend on pollinators, and 87 of the

115 most important food crops depend upon a declining pollinator community. It is a perverse trade-off—farmers are addicted to an insecticide that will eventually destroy farming. Mortality goes beyond pollinators. Pesticides erase springtails, fungi, beetles, ants, mites, bristletails, Symphyla, and other organisms in the soil. Soil and pollinator life resides largely in the hands of chemical companies. The average person has no say or sway.

Insects also disappear due to deforestation, loss of wetlands, lack of wildflowers, and, in the case of the UK, the bulldozing of seventy-five thousand miles of hedgerows that was actively encouraged by the government. Farmland bird populations halved in Europe due to insect loss. Insect-eating birds such as martins, wagtails, pipits, and dozens of other once-common birds are endangered. The food pyramid rests solidly and irrevocably upon a single base—insects.

Ignorance of insect ecosystems underscored Mao Zedong's 1958 mandate for the Chinese people to join the Four Pests Campaign—eliminate rats, mosquitoes, flies, and sparrows. There were chronic grain shortages in China, and sparrows were seen as part of the cause. It was officially estimated that a sparrow could eat two to four pounds of grain per year. People mobilized to join the Four Pests Campaign with utter compliance, as could only happen under draconian communist rule. Every known means to harass and kill sparrows was undertaken. Nests were destroyed, flocks were shot in the air, and drums were beaten in broad areas, making sparrows too scared to land until they fell from the sky dead from exhaustion. Tens of millions of sparrows perished as part of the Great Leap Forward. The sparrow population was virtually eliminated. In 1960, Chinese ornithologist Tso-hsin Cheng explained to Mao's advisors that sparrows ate insects, especially in the summer. It was too late. Sparrows eat seeds, and grain is a seed. They also eat locusts to protect the birds' food source. In other words, sparrows were the Chinese farmers' allies. The locust population exploded in 1960 without any predators to inhibit them. Along with bad weather, grain production collapsed, and a famine of horrific proportions followed. It is estimated that forty-five to seventy-eight million people starved to death. The total death toll in the Second World War was fifty-five million. There was cannibalism, beatings, crime, and murder. The carnage and aftermath remain taboo subjects in China. Chinese students are not informed about it to this day. To rehabilitate avian ecology, China imported 250,000 sparrows from the Soviet Union.

When it comes to planet Earth, we are all amateurs, just like the Krefeld Entomological Society. They are not entomologists. They are priests, schoolteachers, technicians, and enthusiasts. In French, *amateur* means the one who loves. Around the world, thousands of organizations and "amateurs" who love the natural world are acting in diverse, rigorous, and practical ways to bring back habitats for insects and stop their poisoning and destruction. The constructive answer to restoring insect populations is to plant many diverse, colorful, edible, blooming plants that change farmland ecosystems. Conventional farmers are surprised that these techniques make farms more resilient, profitable, and self-sustaining. Restoring farm diversity is championed by Dr. Stefanie Christmann, a plant scientist working directly with farmers around the world. When she first proposed enriching and diversifying farms with prairie strips containing wild grasses, planted fence lines, and verges redolent with flowers, fruits, berries, oilseeds, and trees, she was laughed off the podium at an international farming conference. Since then, her techniques have effectively increased pollinator numbers and the quantity and quality of crops. In semi-arid areas where she specializes, yields of beans and vegetables increased from 177 percent to 561 percent with fewer pests, aphids, and greenflies. In farms with row crops, the technique differs. A quarter of the fields are dedicated to flowering crops such as canola when growing corn, soy, and wheat. Perimeter fencing is planted much like English hedgerows, a wild strip of currants, blackberries, rosemary, salvias, honeysuckle, beech, wildflowers, and crab apples.

Homeowners are changing their back- and front yard plantings to nourish bees, butterflies, and moths. Cities use street medians to create pollinator corridors extending beyond the city limits. Insects are just like us; they want to be safe. By connecting networks of pollinator corridors, insects can avoid poisonous sprays and lands contaminated by Roundup. Farmers are planting prairie strips in croplands, restoring riparian buffers, removing invasive species, and sowing cover crops comprised of dozens of plant varieties to regenerate insect diversity. Volunteers introduce milkweed, natives, and wildflowers onto verges, school grounds, and roadsides. Docents are teaching children about insects at school. Photographers are posting "bugshots" featuring their best insect portraits.

The little things run the world by their intricate interactions with living systems. The climate emergency is not misnamed. The root of the word *emergency* means to arise and bring to light. As mentioned at the outset of the book, global warming is a teaching, an offering, a guide. So, too, is insect collapse. The disappearing invertebrates with their invaluable wings, horns, claws, pollen baskets, thoraxes, mandibles, and antennae bring to light what was ignored—life is here freely, but that does not mean it can be freely taken if we are to thrive. And if possible, stop cutting the lawn. Wild milkweed grows there, the food and birthplace of monarch butterflies. And female fireflies cling to the undisturbed habitat of moist tall grasses and sparkle their abdomens to males who flash in response.

TWELVE

Primeval

For me, the door to the woods is the door to the temple.

MARY OLIVER

Forests preceded human presence on the planet by hundreds of millions of years. Woodlands, swamp forests, and jungles were wild, complex bastions of plant and animal life. Over three hundred million years ago, relatives of today's dragonflies the size of seagulls whirred through the corridors of prehistoric forests while yard-long scorpions and eight-foot, frond-munching millipedes hunted underneath the canopy. Ferns and low-growing club mosses found in today's garden nurseries descend from 130- to 180-foot-tall trees. Extravagant ecosystems accumulated biomass over eons. Due to the lack of fungi and microbes that decompose woody fibers, giant peat bogs accumulated that were transformed over millions of years by heat and pressure into coal. The location of major coal deposits reveals the sites of the most prolific swamp forests. North America has the world's most coal reserves, primarily in Montana, Illinois, West Virginia, Kentucky, and Wyoming. Source vegetation, heat, and age determine whether coal is soft brown lignite or lustrous black anthracite.

Forests comprise 80 percent of all biomass on the land; approximately three trillion trees occupy one third of vegetative land and hold more than one half of organic carbon. Today, they are undergoing a rate of change not seen since the Chicxulub meteorite impacted the Yucatán Peninsula sixty-six million years ago. A rock the size of Mount Everest hit the planet traveling forty thousand miles per hour. The sudden air compression created temperatures hotter than the sun's surface. Sections of the Earth exploded and were propelled into space. It is said that bits of dinosaur bones may be scattered across the moon. Clouds of dust, cyclones of ash, and rains of glass enveloped the Earth from the massive rupture of the mantle, more than twenty miles deep and sixty miles wide. Total darkness descended in many regions, lasting two years. Without photosynthesis, 75 percent of plant life perished. The 180-million-year dynasty of the dinosaurs came to an end. Jay Melosh of Purdue University has modeled the breadth of the Chicxulub impact and believes the vast majority of animals perished, many instantly barbecued. Flowering plants with dormant seeds emerged several years later when the darkening clouds lifted.

No single event is killing off today's forests. Losses are caused by fire, mining, roads, cultivation of palm oil, logging, and the eradication of Indigenous inhabitants and native animals, an overall rate of disintegration unseen since the meteorite. We have a cognitive bias, dismissing plants and trees as inferior to other life-forms. According to Sarah Kaplan, this results in fewer resources being "devoted to the organisms that supply Earth's oxygen, feed its animals, and store more carbon than humanity will emit in 10 years." One in six trees, the largest and longest-lived organisms on the planet, is facing extinction, including the California coastal redwood, the tallest tree in the world. Plant biologist Murphy Westwood of the Morton Arboretum in Illinois mourns, "We're losing species before they even get described."

Rather than ponder complex computer models for future climate impacts, I wanted to know more about previous eras when "green mass" marched north as forests are doing now, a time of warming that exceeded current levels, and thus a glimpse into our possible future. What was the Earth like? Which species benefited and which did not? What life-forms migrated and colonized the northern hemisphere? The northern hemisphere became semitropical during the Eemian period, 115,000 to 130,000 years ago. It was not caused by increased atmospheric carbon, a fact touted by many a climate denier. Carbon levels were roughly the same as at the beginning of the industrial age, 280 ppm. Rather than being a greenhouse gas event, it was a wobble event.

The Earth's axis tilts over a one-hundred-thousand-year cycle named after its discoverer, Serbian scientist Milutin Milankovitch. Various cycles infuse our planet and personal life—seasons, migrations, circadian rhythms, the moon, crop rotations, and even musical rhythms. The Milankovitch cycle is the largest, an elongated axial tilt toward and away from the sun, causing polar ice to expand and disappear, a one-hundred-thousand-year version of our annual seasons, and an accurate long-term climate predictor. The warming periods, which last about fifteen thousand years, are called interglacials. We have been in one for the past ten thousand years. The previous interglacial was called the Eemian, named after the Eem River in the Netherlands. Mollusks there were distinctly different from those in the North Sea in excavated soil beds beside the river. They were dated to be about 118,000 years old.

In 2004, I was invited to visit an Arctic research station in the north of Greenland populated by climate scientists from fourteen countries. The only way to get there is by air. I flew from Kangerlussuaq in southwestern Greenland to the North Greenland Eemian Ice Drilling research station in a C-130 Hercules piloted and crewed by the 109th Airlift Wing of the New York Air National Guard stationed in Scotia, New York. The outpost is in the Northeast Greenland National Park southeast of the Petermann Glacier. A month before, one hundred square miles of the forty-three-mile-long glacier area broke off and became the Petermann Ice Island. The research facility was the farthest I had ever been from a human settlement on land. The pilot knew the area well: "The closest human artifact to our location might be a candy wrapper two hundred fifty kilometers away." On board were the crown princess of Sweden, the crown prince of Norway, and the crown prince of Denmark—Victoria, Haakon, and Frederik, in that order—Nordic royalty who had come to gather a realistic understanding of what their countries face in a warming world. It would seem that the glaciers, ice, and snow in the far north are buffers against rapid climate change. It is the opposite. Temperature increases in far northern and southern latitudes are projected to be three times greater than in temperate zones.

We landed in a virtual whiteout. Out of the blinding snow came staff to greet their royal guests, and I suspect they were just as happy to see the mailbag. In 2007, the ice station began drilling through the ice sheet to retrieve ice core samples from 115,000 to 130,000 years ago. Greenland was 3–5°C warmer at that time, not far from projections of a 2–4°C increase in the coming decades. What happened in the Eemian is happening again, but not in slow motion. The drill site was chosen for the ice's age, depth, and characteristics—8,340 feet to bedrock. Watching the ice cores emerge from the diesel-powered drill barrel in the underground research cave, I could see that they were preceded by a steaming, white, milky liquid that evoked a sci-fi horror movie of my childhood, *The Thing from Another World*, which I saw when I was seven years old. Scientists drilling deep into Arctic ice accidentally rupture an alien spacecraft, releasing a creepy alien who does not like scientists. Here, the "thing" turned out to be overheated coconut oil lubricants used because mineral oils contaminate the data. The core samples revealed past temperatures, impurities in the atmosphere, gas bubbles of the actual atmosphere, and biological material such as pollen. The pollen and isotopes of hydrogen and oxygen found in the cores attested to fifteen thousand years of unusual warmth. Globally, sea levels were sixteen feet higher, and temperatures in Alaska and Northern Europe were 4–5°C higher. Hippopotami wallowed in the Thames River delta, and cave lions feasted on straight-tusked elephants in Germany.

The changes in flora and fauna during the Eemian occurred over thousands of years. The atmospheric changes currently being predicted will occur in decades. To counter the rapid increase in carbon emissions, there are proposals to plant a trillion trees to reduce a portion of our carbon-emitting history. This is especially popular with companies who choose not to reduce their emissions for the time being. Vast planting proposals are couched in percentages of past emissions that could be offset over time. Proponents calculate that half a trillion trees could sequester 25 percent of global emissions. What is generally unstated in the projections is the timescale. Carbon sequestration would be attained many decades from now. When planted on spreadsheets, trees do many good things, but not in reality. Such schemes generally do not consult or collaborate with the traditional landholders where the trees will be planted.

No company can achieve "climate virginity" by putting pine seedlings in the ground. There are wiser voices in the science community. Nature doesn't plant trees; it grows forests, resilient communities of trees, plants, and animals. Humanity's total yearly emissions of carbon is about 11 billion tons. However, the net annual increase in elemental carbon in the atmosphere is about 5.4 billion tons because land, plants, and oceans sequester 5.8 billion tons. Forests capture the most carbon dioxide on land, and existing mature, primary forests are responsible for the great majority. Until recently, it had been assumed that the older trees in ancient forests sequestered carbon marginally, if at all. It is the other way around. Primeval forests accumulate significant amounts of carbon toward the end of their long lives. Planting trees on bare land is like feeding a bird in a cage. Forests have deep mycological connections and symbiotic relationships with soils rich in fungi that function as the other half of a tree. Without a living underground community, tree plantations do not sequester carbon effectively. Protecting existing forests would have far more impact between now and 2100 than newly planted forests.

The most important terrestrial ecosystems for carbon are the five megaforests on the planet: the boreal forests of Canada and the Russian taiga; the Amazon; the Congo Basin; Papua New Guinea; and Indonesia, including Borneo. The capacity of megaforests to store and receive carbon is fostered by cultural diversity. There are over 1,000 languages spoken in New Guinea, 300 in the Amazon, 653 in Indonesia, and 170 in Borneo, most of which contain teachings, stories, and knowledge on how to live within and sustain the forest. The Momo people go back at least fifty thousand years in Western New Guinea. Their forest homes were intact until recently. The unbroken portions of megaforests result from cultural lineages that see the forest as family and kin. With kinship comes obligation, loyalty, and respect.

Megaforests are characterized by being roadless. Biologist Tom Lovejoy, who coined the term "biological diversity," did pioneering research in the Amazon. It revealed how forest fragmentation caused by roads and pasture clearings was followed by a drastic reduction in species diversity and forest health. Lovejoy set out to determine the minimum critical size of land required to maintain ecological diversity. There had been a debate about whether biodiversity could be protected by conserving several smaller areas or whether protection required large, unbroken landscapes. His research demonstrated that it was indisputably the latter. Intact forests are cooler in temperature, make rain, amplify diversity, and offer greater abundance for inhabitants. Broken forestlands are drier; trees are subject to stronger winds; fire is more likely; settlers may squat and carve out subsistence plots, and species vanish. Easy access causes animals crucial to dispersing indigestible seeds, such as tapirs and agoutis, to be hunted for bushmeat.

There is nothing permanent about forests; they come and go. The planet has oscillated in a one-hundred-thousand-year cadence for the past three million years, alternating between ages of glaciation and warming. During the warming periods, trees move north. During ice ages, they retreat. Ben Rawlence puts it this way: "Time-lapse photography of geological time would show a sheet of ice descending and retreating in a rhythmic pattern, and a green mass of forest rising toward the North Pole then falling again, like breath." That green mass is communities of pine, larch, spruce, fir, shrubs, mosses, and lichen, creating mysterious habitats replete with quagmires, marshes, and bogs, arctic swamps populated by stubby trees draped in canopies of black lichen.

Megaforests are the wildest places on earth and the most

diverse concerning speciation. The North American boreal is the largest unbroken expanse of contiguous forest, peatlands, and wetlands, interlaced by streams, rivers, lakes, and ponds, forming an area of approximately 1.2 billion acres. One to three billion birds fly to the North American boreal for their summer refugia, migrating from as far away as Patagonia. In the fall, three to five billion birds and hatchlings fly back to wintering sites. These include birds seen in backyards, parks, fields, and forests—warblers, sparrows, ducks, waxwings, and ravens, and endangered species such as the remaining four hundred whooping cranes.

The boreal forests are the lair of wolves, grizzlies, wood bison, caribou, moose, and a profusion of small carnivorous mammals—lynx, martens, minks, stoats, sables, wolverines, badgers, and weasels. Summers are cool and short, and winters are long and cold. In many areas, soils are thin, sandy, and toxically acidic due to the constant deposition of needles, resins, oils, and chemicals from the trees. In areas where light can penetrate, there are fairy-tale patches of salmonberries, blueberries, and red and black currants. In bogs and fens, carnivorous sundews and pitcher plants trap and digest unsuspecting insects and spiders.

The conifers that dominate the boreal are dark green to maximize light absorption. They form perfect pyramidal cones to shed the heavy winter snow and produce antifreeze resins in the needles to keep them from freezing solid. The boreal has the highest carbon density of any region on earth, with more carbon below the ground than intact tropical forests have above the ground. The damp, cold conditions in the boreal prolong decay and create carbon-rich bogs and peatlands. When boreal forests are harvested and clear-cut, the disturbance dries out the soil, which releases carbon emissions greater than the loss of the trees. If half of the boreal forest and its carbon stock is lost, carbon dioxide in the atmosphere will reach 600 ppm compared to 425 ppm at this writing.

In Canada, Scandinavia, and Russia, the boreal is home to more than six hundred Indigenous communities who know the land, forests, and waters like no other. At the boreal latitude, there is more carbon under the lakes, in the woods, and throughout the peatlands than in the atmosphere. Today, the boreal in Canada is becoming fragmented, picked to pieces by mining and industrial forestry. Companies are destroying pine trees for toilet paper. Zoë Schlanger writes, "What did it take for that tree to live through those years, make thousands of leaves each spring, store sugars through the winter, turn light and water into layers and layers of wood? It is hard to underestimate the drama of being a tree, or any plant. Every [plant] is an unimaginable feat of luck and ingenuity. Once you know that, you can't unknow it. A new moral pocket has opened in your mind."

In their book *Ever Green*, John Reid and Thomas Lovejoy describe the biological scope of primary forests: "They have predation, pollination, seed spreading, and procreation all happening naturally and in profusion. They have troops, colonies, packs, and pecking orders: microfauna, megafauna, intrepid migrants, and entrenched residents. Harpy eagles eat spider monkeys, grizzlies eat salmon, tree snakes eat tree frogs, pitcher plants eat ants, and ants farm fungus." Tropical megaforests were once called jungles, a word derived from the Hindi *jangal*, which meant an impenetrable forest unfit for human beings that required intense, competitive struggle to survive. Today, tropical forests are portrayed as ancient environments with natural hazards, venomous snakes, lurking predators, and thin soils. On the contrary, the inhabitants of tropical forests, such as the Achuar, believe they once had the least stressful life compared to other people on the planet. Their stress now arises from the incursions and destruction of ancient ancestral homes by mining, drilling, agriculture, dams, and logging.

Protecting megaforests is five to seven times less costly than reducing emissions or planting new forests. In other words, being effective regarding the atmosphere and living world is the least expensive. However, that would be only one measure of cost, a colonial way of seeing the value of megaforests. An inclusive view extends directly to culture. For thousands of years, human beings have thrived in these environments. Africa contains a sixth of the planet's forests, and 70 percent of the population depends on forests for their livelihood and sustenance. Indigenous inhabitants have continuously modified the landscape of tropical areas and created wild forest farms. Norman Myers, Oxford scholar and famed British conservationist, recounted visiting the lowland rainforests that envelop Borneo because he wished to see an "untouched, virgin forest." Borneo is the third largest island in the world, twice the size of Germany. The tropical broadleaf forests there contain an estimated fifteen thousand species of plants, more than all of Africa. There are towering stands of dipterocarps, including ebony, ironwood, and strangler figs. He journeyed far into the forest with an ethnobotanist to see what a forty-thousand-year-old forest looked like. They stopped deep in the woods and remained there for hours, slowly rotating their focus as the botanist identified the trees, shrubs, and vines before them. When the day was over, it was evident that the spectacular diversity arrayed was not an untouched, ancient forest. They saw the result of interactions by Indigenous peoples who resided in their homes for thousands of years. When towering trees weighing seventy thousand pounds succumb to age and fall to the forest floor, other trees are taken down, and swaths of sunlight flood the forest floor. In those areas, people plant seedings and cuttings that will provide the medicines, food, fiber, and wood needed for the future. To quote Reid and Lovejoy: "For modern humanity to keep the megaforests, and with them the one planet we know of that has any forests, we need to care for the world as if it is family. We must attempt a grammar in which subject and object, people and everything else, are the same. In a material and evolutionary sense, they absolutely are."

THIRTEEN

Dark Earth

*When you are standing on the ground, you are
standing on the roof of another world.*

JILL CLAPPERTON

Underfoot is the most complex living system on Earth.
Dark, slate-gray, sepia soils hold entangled menageries of
life-forms, most of which science cannot identify and has never
seen. This intricate womb of loam, clay, silt, and marl feeds and
constructs the endless complexity of our world. Soil is an artifact of a lineage going back eons made of fungi, microbes, insects, and minerals, a rich brew of elements that orchestrate the
cycles of life from decay to sentience. Describing the subterranean world of soil requires language lying somewhere between
science and poetry. How could a dirt clod hold wonders? How
could a teaspoon of soil contain six miles of mycelia? Science
can analyze and sequence organisms found in soil, but it cannot
make soil. People can create the conditions that engender fertile

dark soils and have been doing so for thousands of years. However, only the inhabitants of soil create soil. Millions of years of coevolution reside in and upon the earth, forming, in the words of Peter McCoy, the "skin of the world, tattooed with the legacies of its inhabitants." There is a growing movement to rewild ecosystems and bring back lost or declining species. Soil is the wildest organism of all. Unless we breathe life back into the ground, the web of life will fail. Conversely, soil will atrophy and perish without the myriad creatures that create it.

Billions of microbes and fungi of indecipherable complexity reside in the soil. When removed to the lab to be cultured, most microbes perish before they can be identified. Soil fungi comprise vast labyrinthine networks that cannot be analyzed when cut into microscopic snippets. Untethered fungal nuclei float alone in tubes one cell thick. How to study a system where there are no individuals? The interaction between microbes, fungi, plant roots, and insects makes up a world upon which humanity depends. Science is accustomed to understanding the whole by analyzing the components. In the case of soil, it may be the other way around: to comprehend what is happening below, the living mantle of the Earth requires an overarching understanding. The term *Mother Earth* was not coined as a sentimental kindness. It expresses the primordial truth of the origin of life.

Ninety percent of insects and invertebrates spend some or most of their life on or within the soil. We generally ignore them unless they affect us, such as termites. Insects gobble leaves, roots, fungi, and each other, fertilizing the soil with their excretions. Their activity benefits the soil but not all food crops. Digger bees raise their families in well-drained soil where they store nectar and flowers. Bright-eyed cicadas, with their yellow wings and bulbous green bodies, can be tucked away in tree roots for up to twenty-five years. Their neighbors include cone-heads, garden darts, moss mites, mud daubers, wart-biters, springtails, roly-polies, crested millipedes, yellow-shouldered ladybirds, cockchafers, and more.

Ninety-five percent of insects are beneficial to soil and plants. Several species of ground beetles are treasured. They feed on pests, including the larvae of leafminers, cutworms, and aphids, and eat weed seeds, seasonally alternating their diet from insects to seeds and back again, creating biological pest controls that are a farmer's best friend. In return, some farmers create strips of tallgrass in their fields known as beetle banks, where they are safe from predation and can winter over with food and cover.

If there is a heroine in the soil, the earthworm wins. It slithers along, eating decaying roots, leaves, manure, nematodes, bacteria, microbes, and fungi twenty-four hours a day. What emerges out of the digestive system of the "planet's greatest alchemists" is vermicast, considered the preeminent fertilizer on the planet. Excretions contain minerals, enzymes, microbes, and bioavailable nutrients for plant roots. Agroecologist Nicole Masters writes, "Vermicast improves seed germination, plant health, and production, beyond the wildest dreams of synthetic [fertilizers]—all at much less cost to producers and the environment. Some varieties work on the top surface of the soil, others lower down. Anecic worms go six feet down, bringing minerals to the top and organic material to the bottom."

Charles Darwin first brought earthworms to the world's attention as a keystone species. Despite an onslaught of ridicule, Darwin championed the earthworm, forthrightly stating that no other species had "played such an important part in the history of the world as these lowly organized creatures." Although schooled as a geologist, Darwin keenly observed that species evolved by adapting to their environment rather than being rigidly fixed in form and function, as was put forth by religious dogma. Geologist Charles Lyell enlarged the scope of evolution by showing how natural causes shaped geological changes over millions of years. Given Lyell's insights, Darwin's uncle Josiah Wedgwood suggested he investigate how the earth emerged and subsided over shorter periods than seen in sedimentary formations. To do this, Darwin worked closely with his children to study earthworms that tunneled through the soil at his twenty-acre Georgian manor in Kent. At dawn, the family would troop outside to measure the volume of worm castings upon the surface in the front gardens; more soil on top meant less soil underneath. Darwin estimated how activity created layers of subsidence employing an unusual metric: a weighty circular "worm stone" was placed on the lawn, and they watched it slowly sink more deeply into the earth.

When it comes to soil engineers, dung beetles may be the most beguiling. The most notable of the dung beetles live in the African savanna. They gather animal manure and roll it into balls up to ten times their weight and size. In one night, the dung beetle can bury two hundred times its weight in ordure. With their front legs atop for stabilization, beetles push with their hind legs and roll dung into their breeding chambers. To give dimension to the strength of a dung beetle, imagine pushing a fifteen-hundred-pound ball of excrement half a mile along uneven ground. Dung is all they eat. Egyptians revered dung beetles as scarabs, symbols of birth, life, death, and regeneration. Stone carvings of scarabs were placed on the hearts of the deceased in tombs. To reproduce, dung beetles create brood holes six to eight inches deep, roll in the ball, and lay their eggs inside the dung, providing food for the larvae. Dung beetles improve soil structure, bury seeds, and recycle plant nutrients. They are so effective at cleaning the lands of manure, they are bred and flown to different parts of the world to remedy pasture infertility, where animal droppings remain intact and become breeding grounds for parasitic flies, and into urban areas where there is a buildup of dog poop in parks and greenways. The nocturnal dung beetle, a custodian of terra firma, has a near-mystical sense of orientation. On clear nights, beetles sense the Milky Way using polarized starlight to navigate and orient themselves. When the moon outshines the stars, the moon becomes the compass. Found on every continent except Antarctica, the dung beetle is disappearing as livestock is removed from the land or because antibiotics and veterinary medicines are poisoning the manure they find. The dung beetle creates soil; industrial agriculture destroys soil. According to the United Nations Environment Programme, twenty-four billion metric tons of topsoil is lost annually to erosion, over six thousand pounds per person.

Ants, it is said, rule the world and permeate the soil. According to E. O. Wilson, were it not for the presence of *Homo sapiens*, Earth would be the planet of the ants. Before we arrived, they did just that: they ran the world and managed the flora and microfauna decisively and generatively. Single Argentine ant colonies are larger than Texas. There are 2.5 million ants for each human being, and the dominant, no-questions-asked majority are female. Male ants are ciphers with wings and colossal genitalia that have one role in life: to inseminate virgin queens. Described as flying sperm missiles, they die within a week of their one task in life, whereas queens can live for one or more decades. The rule and role of ants in soil are crucial. Ants eat leaves, sap, aphids, fungi, animals, nectar, larvae, lizards, amphibians, Symphyla, and their own dead and injured. They aerate the soil in vast tunnels, depositing nutrients. Leafcutter ants march single file carrying colorful fragments of leaves, fruits, and flowers into vast underground nests where they cultivate fungus gardens made of freshly chewed vegetation. Fungus is their only food. The subterranean mounds of their underground nests can span one hundred feet and contain millions of individuals. Excavated soils from one nest can weigh fifty tons or more.

In a cup of soil, billions of creatures, microbes, and organisms feed upon each other in one extensive buffet. Arrays of nematodes join the feast and regulate the microbial population by what they choose to eat. Most people have never heard of nematodes. There are over one hundred million species, an estimated sixty billion nematodes for each human being, representing 80 percent of individual animals on the planet. Nematodes permeate layers of the lithosphere down to twelve thousand feet. Under the microscope, they appear like tiny strands of curly hair. In the environment, they redefine the word *ubiquity*; they thrive in the Arctic to the bottom of the ocean. They also live within the entire animal kingdom (thirty-five species are endemic to the human body).

Nicole Masters compares the soil's digestive process to our own. We chew food into saliva-infused boluses that are broken down by gut bacteria, which create enzymes that make nutrients available to our bloodstream. Over the past two decades, the understanding of the human gut has been transformed from being analogous to a sewer into a biome called our second brain. The soil biome encompasses a similar digestive process and is undergoing a similar sea change in understanding. Seen as a medium into which soluble fertilizers, plows, pesticides, and fungicides can be inserted, healthy soils are now recognized as a living ecosystem, the source of vibrancy, nutrient density, minerals, water, and resiliency. The human gut microbiota is crucial to physical and mental health and requires a diverse diet of nourishing, unprocessed food for maximum well-being. Ultra-processed food, sugar, additives, artificial sweeteners, preservatives, over-supplementation, rancid fats, alcohol, and drugs play havoc with the human gut. Analogous to human digestive dysfunction in the soil are herbicides, neonicotinoids, carbamates, algicides, soluble fertilizers, nematicides, insecticides, and more—a long list of industrial "junk-food" products sold to farmers to maintain yields and fend off predation, but which destroy the life of the soil. The unending use of chemicals and toxins destroys complexity and the fecundity created by the incalculable number of interactions between fungi, soil creatures, and microbes.

The world is at an inflection point with respect to how humanity will produce food. Modern farm techniques confuse extraction with management. Despite the warnings, monocultures are expanding, sowing higher-yielding seeds with identical genetics that require more pesticides and fertilizers. Crops are not being developed for you and me. They are engineered to make money, to manufacture inexpensive starches and sugars for ultra-processed food, to provide feed for caged pigs, cattle, and chickens. Our knowledge about agriculture is based on observing and analyzing denatured soils low in biodiversity, bereft of fungi, and meager in carbon. There is a good chance that most farmers have never seen fully functioning soil. Depleted soils are considered normal. To personally understand the breadth and depth of what has happened to soil, stand on it. If it does not make way for your footprint, it is not healthy. It has become compressed. The spaces for water and air have disappeared. The lack of structural integrity results in the lack of habitat for microbial and fungal life. It is a positive feedback loop: less life, less structure. Rather than being the midpoint between sand and brick, living soil is remarkable to the eyes and nose. You should be able to dig it with your hand. Hold soft, dark, crumbly soil in the sunlight and examine it closely. It should sparkle and feel moist. Hopefully, there will be earthworms and small arthropods. Its ebony shades emit a fragrance that is unlike any other scent. It has a name, petrichor or geosmin, and is said to be created by essential oils emitted by plants that are absorbed by sand and rocks. Your nose and body know it, the smell of life, and it is good. The life that created the soil was a plant, a life created by the energy of the sun. Soil is the link between the light of the sun and the gleam of an eye. This is not meant to sound poetic. It is true.

Farmers were taught a system of agronomy that seemed remarkable at the outset. Direct chemical applications combined with mechanically tilled soil increased yield, income, and security. The surface application of macronutrients—nitrogen, potassium, and phosphorus—meant plant roots did not have to go far for their basic needs. However, the plant was less likely to obtain the complex mineral nutrients found deeper in the ground. Modern tillage techniques pulverize soil, release carbon into the air, and destroy macropores that provide oxygen and water to the soil's biological communities. Rinse and repeat annually, and the soil becomes dirt, not a living system. It is dependent on "inputs," the agricultural term for chemicals.

The Green Revolution in the 1960s created higher-yielding varieties of rice and dwarf wheat that required greater amounts of pesticides, herbicides, and synthetic fertilizers. At that time, there was a widely held belief that chemical agriculture could prevent hunger in the world. No one contests that these changes created more food. It was called the "green" revolution as a way to contrast it to the Russian-inspired "red" communist revolution that was increasingly popular in countries dominated by hunger and poverty. The US goal was to end hunger and create food security to counter that threat. The Green Revolution was a triumph that presented the world with a massive slow-motion hangover. Monocultural agriculture (wheat, corn, and soy) requires more machinery, fossil fuels, poisons, tillage, and plowing. The uptake and reliance on synthetic fertilizers, pesticides, and fungicides grew at twice the rate of increased yields. During this era, the food system became the single largest source of greenhouse gases, nearly one third of global emissions. The revolution degraded the life of the soil in order to extract more food. One third of arable land has been lost to industrial agriculture in the past four decades, a rate of soil erosion one hundred to one thousand times greater than natural erosion rates. When the high-yield, high-input victories are totaled up, the ecological costs are not mentioned. Farmers did not benefit from the Green Revolution's higher yields. Large corporations benefited from lower commodity prices. Soft drink companies save money using high-fructose corn sweeteners, and Iowa became the corn capital of the world, with 64 percent of the crop dedicated to ethanol production for cars and trucks. It requires more fossil fuel energy to create the energy produced by corn-based ethanol combustion, a net overall loss and major contributor to increased greenhouse gases, chemical and water pollution, oceanic dead zones, and loss of pollinators. It is the "clean fuel" that airlines are lining up to purchase. Farm families that depend on commodity farming are more financially insecure today than ever before. They are indebted, stressed, and work in toxic environments. Some farmers will not eat what they grow but cultivate an organic garden for their family. Today, three billion people cannot afford a healthy diet, equal to the total world population at the start of the Green Revolution.

Rather than being reservoirs of life, industrialized soils are closer to dry lakebeds. Climate change and conventional agriculture are in a vicious cycle. Each degrades the other. In the foreword for *Agricultural Testament*, Wendell Berry points out that chemical agriculture has never known what it is *doing* because it does not know what it is *undoing*. Global agencies that address hunger, food, agriculture, desertification, and deforestation are on high alert, overwhelmed by the rate of land disintegration. Restoring the soil is about active processes that allow inherent regenerative processes to return to the land. The interactions between bacteria, microbes, viruses, fungi, ants, earthworms, insects, and nematodes are incalculable even in a square foot of soil. If the little things run the world, the littlest of all may hold the greatest influence. Single-celled microbes can have one hundred thousand sensors on their cell wall that detect and respond to the immediate environment. There are over a billion microbes in a teaspoon of soil. The microbial complexity of soil is incalculable. So too are the interactions of mycelia, nematodes, light, rain, roots, and flocks of dung beetles flying off in the night to find new cow patties. The smallest beings determine the texture, fertility, composition, moisture, and nutrients in the soil.

Soil is the dance and flow of carbon under the surface of the earth. There is a soundtrack for the dance. Over the past two decades, scientists have inserted microphones into the soil and turned up the volume. There are thrums, chirrs, chirps, trills, and rasps. There are rustles like the soft, muffled sound of dry leaves and water moving through pores. One can hear clicking sounds similar to vocalizations employed by sperm whales and the San people of Botswana and Namibia. Mole rats are heard banging their heads against their tunnel walls. A researcher described the sound of soil as the creaking reverberation of a large tree in the wind. The combined utterances in rich, diverse soils may collectively sound like pieces of fine sandpaper rubbed together. Even plant roots growing through the soil make audible sounds. Watch a robin in the spring cock its head to the ground, a hungry bird listening for larvae and earthworms. Scientists who research the sounds of soil note that industrial farms employing machinery and chemicals are "eerily quiet."

Because of its momentum and dominance, asking industrial agriculture to become ecological is akin to asking a locomotive to make a U-turn. Commercial agriculture strangles and constricts the flow of carbon, restricting the involvement of lifeforms inherent to healthy soil. The passage to regenerative farming is through the soil, not a laboratory. There is no other way in. What prevents the implementation of ecological agriculture is an industry comprised of some the biggest chemical companies in the world. They are in a dilemma because the means to regenerate soil cannot be sold in a can.

The conventional food industry presumes that industrial agriculture is here to stay, soon to be joined by vertical farms, lab-grown chicken, and vat-cultured meat. Artificial initiatives will move the food system further away from the living system of the soil toward an ever-greater concentration of production. Corporate food production envisions farms without farmers, overriding life-giving agricultural techniques because regenerative, land-based methods do not lend themselves to corporate-scale profits. Restorative farming techniques and practices have numerous designations: agroecology, ecological, organic, biodynamic, and regenerative encompass most methods. These farming technologies employ the breadth of plant/soil/entomological/fungal relationships in specific landscapes. These techniques are more science-based than industrial methods. Sophisticated, biological, nontoxic ingredients are employed that break down without harm. There are farming techniques that culture the soil's innate microbial population in small vats, liquid ferments that might be called soil kefir, which are then returned to the soil. The land tells the farmer what it wants.

Because soils and grasslands are degraded and dehydrated, there is a growing cadre of land doctors who heal the land: Judith Schwartz, Nicole Masters, Chris Henngeler, John Liu, Brock Dolman, Christine Jones, Charlie and Tanya Massy, Hui-Chun Su, Dianne and Ian Haggerty, and hundreds more. Using herbivores, prairie strips, key lining, crimpers, gully stuffing, microbial brews, complex rotations, fire ecology, and a bevy of other tools and techniques, they treat land degradation and soil pathology. Abraham Lincoln coined the term that applies to modern agriculture: "bass-ackwards." Plant breeders have spent a century creating seed varieties that grow in impoverished soils when we should have focused on soil restoration.

Regenerative agriculture has a simple maxim: create more life, above and below, step-by-step. Soil tests reveal much about the soil, but so do the weeds. Weeds that bedevil and invade reveal imbalances in the soil below, a response to underlying soil dynamics. Farmers employing regenerative grazing techniques, such as the famed thirty-thousand-acre wheat and sheep farm in Western Australia cultivated by the Haggertys, report the reappearance of beneficial native plants that had disappeared decades before. The community of farmers in the world who practice ecological farming are being taught by their land, soil, and crops. The knowledge is shared among colleagues and is gradually being taken up by agricultural schools.

The tallgrass prairies of North America stretched from Canada to Oklahoma and across the Midwest. The 240 million acres were home to bison, elk, deer, and antelopes. They were populated by forbs, bluestems, Indian grass, switchgrass, and other perennials that grew upward of six feet with roots five to fifteen feet deep. The soils formed a sticky, carbon-rich substance known as glomalin. It glued the soils together, sequestering more carbon than any forest on Earth. The extraordinary accumulation of carbon created the most fertile soil in the world, called mollisol. As Nicole Masters points out, a fertilizer truck did not follow the sixty million bison around to create the richest, darkest, and deepest topsoil on the planet.

FOURTEEN

Untranslated World

*Take a journey beyond the city's dazzling skies
and allow your eyes to adjust to the darkness.
Watch the animals come out from their hiding
places, their eyes glittering, and their silhouettes
passing by. Smell how the plants' scents change,
hear how the new sounds take over.*

JOHAN EKLÖF

The most important farm in the world doesn't grow food. In
1999, Charlie Burrell and Isabella Tree, owners of a thirty-
five-hundred-acre money-losing farm known as the Knepp Estate, situated in Sussex, England, threw in the towel. It came
down to the clay soil. There are dozens of insulting words for
Sussex clay, including *slub, gawm, gubber, sleech,* and *pug.* The
wet, winter muck once swallowed a horse and rider to their
eyeballs. Come July, the clay turned to cement. Sometimes, six
months would pass before they could work the land. They strug-
gled to turn the economics around. Intensified agricultural

methods were deployed: new machinery, expensive fertilizers, clever pesticides, and state-of-the-art milking parlors. Though farm yields improved, and the dairy was rated one of the top ten in the country, the Sussex clay prevailed. Losses mounted. In 2000, they made a decision that has reverberated throughout the conservation world since: they let the farm go wild.

Both Tree and Burrell are naturalists and scholars; Isabella is an award-winning author and conservationist. They traveled several times to the African bush where Charlie was raised. Across the continent were unconstrained landscapes redolent in wildlife—no fences, machines, roads, or agriculture. The contrast between the barren, industrial-farmed lowlands of the UK and the Serengeti plains teeming with diverse wildlife was stark. It led Charlie to wonder: Could they imitate what they saw in Africa and give natural processes free rein at Knepp? Would freely grazing, undomesticated animals create habitats and wildlife savannas across the estate? Neighbors, conservationists, and local authorities were not enthusiastic about the idea.

They were inspired by the work of biologist Frans Vera, a Dutch conservationist, who pioneered a controversial theory of landscape evolution in Europe. It has been long assumed that pristine, old-growth forests are vital to protecting shrinking biodiversity. Vera proposed that "pristine nature" consisted of open woodland pastures created by wild grazing animals, not the fairy-tale forests of Snow White and the Brothers Grimm. Rather than protecting old-growth forests, Vera proposed that grazing mammals should be returned to woodlands, grasslands, and wetlands. He believed the symbiosis between pastures and forested lands results in maximum biological diversity and that the presence of grazing animals has always been the driver of habitat creation. Without free-ranging animals, "you have impoverished, static, monotonous habitats with declining species. It's the reason so many efforts at conservation are failing." At the time, Vera's thinking was heretical, and it remains so for many conservationists.

In 2000, Charlie and Isabella shut the farm and sold off the cattle, cows, and machinery. They brought in animals closely related to species that once populated the UK and Europe. After removing most of the interior fencing and reinforcing the perimeter fences, they introduced fallow deer, Old English longhorns, Exmoor ponies, and Tamworth sows with piglets. The quartet of animals were the close relatives to the extinct aurochs, the tarpan (the original wild horse of Europe last seen in 1887), the wisent (European bison lost in the 1920s), elk (called moose in America), and wild boar. The animals were given free range. There was no supplementary feed or intervention. Beavers and red deer were added later. Neighbors and the government would not countenance apex predators like the wolf, wolverine, or lynx, so an ongoing culling process was instituted as herds built up.

Knepp's unexpected outcome is brilliantly retold in *Wilding* by Isabella Tree, a number-one bestseller in the UK, and the more recent *The Book of Wilding*, written by both Tree and Burrell. Knepp became a stunning testament to Charlie's Africa and Vera's observations of Europe. As the estate changed, significant numbers of species showed up. For Isabella, the most precious arrival was the soft, soothing sound of the turtle dove, rarely heard today in the UK. From a population of 250,000 in the 1960s, there are an estimated 5,000 remaining, with less than two hundred pairs in the county where Knepp is situated. More than a million are shot and killed annually in Europe by hunters during their springtime migration. The loss in numbers in the UK is caused by the deracination of the English countryside by weedkillers, pesticides, plowing, and the absence of native plants. Many other farmland birds are disappearing across the country for lack of food and safe nesting places; quails, lapwings, skylarks, linnets, yellowhammers, and tree sparrows need insects, not pesticides.

The secret to Knepp was getting out of the way and putting nature in charge of itself. Isabella writes, "Rewilding—giving nature the space and opportunity to express itself—is largely a leap of faith. It involves surrendering preconceptions and simply sitting back and observing what happens." On the train to Knepp, I looked out the window at pastures with the telltale bluish-green appearance produced by chemical nitrogen fertilizers. Hedgerows were rare. When I arrived at Knepp, I was surrounded by a scent—the indescribable fragrance of soils, grasses, flowers, meadows, and woodlands. In 2019, an introduced white stork nested on one of the castle turrets. It was the first time this had occurred since 1414, before there was a Britain. When I visited in 2023, there were more than twenty nests. At least half the flock was wheeling overhead, a miniature murmuration effortlessly circling the updraft. Calling Knepp an ark is a banal metaphor, but that is what it is. New passengers arrive daily in flocks, on paws, on the wing, clambering and fluttering in the oaks.

Not only did new bird species arrive, but those already resident increased their numbers dramatically: fieldfare thrush, woodcock, and lesser redpolls, cabaret as they are known in France. Add skylarks and woodlarks, the gadwall duck, and the winnowing sound of jack snipes in flight. Ravens returned after a one-hundred-year absence. Bats skimmed the lakes, including the micro pipistrelle that can fit into a matchbox. In the UK, moths have declined 88 percent in the last fifty years. At Knepp, seventy-six new species have been spotted. Egrets, bitterns, scaup ducks, and green sandpipers joined a single Himalayan bar-headed goose that had escaped from a private zoo. Other geese include the graylag and Egyptian. The butterfly population exploded in number and variety: pearl-bordered, marsh, and high brown fritillaries, gatekeepers, ringlets, meadow browns, marbled whites, small and Essex skippers, common blue and small tortoiseshell, and the rare and much sought-after purple emperor. Nightingales trill, chatter, and sing their flutelike whistles throughout the summer nights. Green woodpeckers gobble meadow ants, insectivorous whitethroats scratch out their nasal yowl, and the piercing cry of cuckoos is heard. Predators arrived to dine: the raptors, ravens, peregrine falcons, eagles, short-eared owls, stoats, weasels, and polecats.

Even though Knepp is surrounded by chemical agriculture, something remarkable is happening with the land. Charlie and Isabella's daughter Nancy studies wildland carbon capture rates for her PhD at Oxford. A recent measure of soil carbon in rewilded grassland at Knepp revealed an annual sequestration rate of 3.4 to 4.8 tons per hectare. An email from Annie Leeson, the CEO of Agricarbon, the carbon testing company, stated that she hadn't "seen such a clear demonstration on anything like this scale, or with this level of evidence, anywhere in the world."

What began in 2000 as an experiment in farmland restoration has become a signpost to the world. According to Isabella and Charlie, they "had no idea Knepp would end up a focal point for many of today's most pressing problems: climate change, soil restoration, food quality and security, crop pollination, carbon sequestration, water resources and purification, flood mitigation, animal welfare, and human health." Knepp reveals the essential path to a biological transition of the world. It doesn't mean abandoning farms. It means that the planet's regeneration is at our feet—animals, soil, and wildness. And it demonstrates a fundamental and inspiring maxim: nature rebounds with astonishing speed.

The Knepp thesis was demonstrably proven in Romania, where European wood bison, known as wisent, were reintroduced into the Southern Carpathian Mountains in 2014. As in America, the Eastern European bison were hunted nearly to extinction. Through captive breeding and rewilding, herds have increased and now number in the thousands. In Romania's Tarcu Mountains, a free-roaming herd of 170 has been reinstated. The rate of ecosystem regeneration is remarkable. Following the teachings of Frans Vera, the bison were placed in a grassland and forest ecosystem. The grazing bison recycled nutrients and scattered seeds throughout their range. Recent analyses of the herd's impact conducted by Oswald Schmitz of Yale University showed the bison capturing sixty-one thousand tons of carbon per year, equivalent to the emissions of forty-three thousand cars. The longtime absence of bison resulted in massive amounts of carbon being released from plowed-up grasslands; their return brought about a rapid reversal of the losses. In the heated debates about climate, grazing animals are generally referred to as problems, because they provide meat and damage rangelands. The Knepp Estate in Sussex and the return of bison to the Tarcu Mountains tell a different story.

Botanist Robin Wall Kimmerer described a plant scientist who hired a young Indigenous guide to gain local knowledge of a rainforest. As they entered the thick and varied growth, the youth described one species after another, its history, name, and usage. His breadth of knowledge was a complete surprise to the scientist, who complimented him on his depth of understanding. The youth accepted the praise but replied with downcast eyes, "Yes, I have learned the names . . . but I have yet to learn their songs." Author and physician Siddhartha Mukherjee referred to the Kimmerer story in *The Song of the Cell.* "What the young man laments is that he hasn't learned the interconnectedness of the individual inhabitants of the rain forest—their ecology, interdependence—how the forests act and live as a whole . . . the songs that move between the trees."

Mukherjee was questioning the premise of atomism, a view of the world that goes back to Antonie van Leeuwenhoek's discoveries of single-cell organisms in 1674. Atomism is the belief that the world can be understood by studying its most minuscule particles, cells, or atoms. It is the foundation of modern medicine and the basis for studies of soil, plants, and animals. As science accumulates unfathomable amounts of information about the bits, parts, and pieces of our bodies and nature, we can easily miss the song, the unheard symphony, the untranslatable world of wilderness. A song connects the contours of planetary life into an intricate, immeasurable tracery of beauty and truth. For Kimmerer, it is the capacity to hear the suffering of people, plants, and animals with one's heart. For marine ecologist Monica Gagliano, it is the difference between the thinking world and the feeling world.

The linchpin of planetary life is animals. We are one of them. The songs that move between the plants, fauna, forests, and fungi are intelligence. It would take forty-five million years to count the number of animals that live on Earth. Contrast that to the average number of animals we see daily. The most likely touch points are either pets or animals cut into parts at the meat counter. Otherwise, we have little or no contact with the 3.4 trillion birds, mammals, reptiles, insects, amphibians, and fish. While it is fair to say that animals prefer it this way, they are losing their homes from Antarctica to Alaska and everywhere in between. As I write, an emerging corporate concern has arisen called "nature risk." After centuries of exponential rapaciousness, it is unsettling to witness how long it took for "nature" to come to the fore in the business world. However, the concerns remain trivial. Impacts include what will happen to data centers because of their dependency on shrinking aquifers, or the plight of supermarket strawberries for lack of bumblebees.

The commercial world seeks clarity, measurement, and predictability. It is hungry for permanent solutions to the climate crisis. Pema Chödrön describes us as people in a collapsing boat trying to hold on to the water. The dynamic, fluid, natural flow of planetary life does not align with our need for certainty and permanence.

Fortunately, numerous nonprofit organizations address wildlife in all its contours. Scientists and activists comb the Earth, count populations, catalog species, restore habitats, mourn the losses, and call out corporations, agriculture, and government agencies that destroy the environment. Their work is heroic and suffused with grief. As with climate, people say they care about nature, especially when portrayed as cute and cuddly, but do little or nothing to stop the senseless losses. Is this a people issue or a communication problem? Populations of wildlife have declined 73 percent since 1970. The 1992 United Nations Convention on Biological Diversity declared that biodiversity should be conserved for the "sustainable use of its components and the fair and equitable sharing of the benefits arising out of the utilization of genetic resources." It is not surprising that 196 countries ratified it. The treaty is an objectifying and extractive view of the flow of life, the flow of carbon. Vast numbers of living beings have disappeared since the treaty was signed—an accelerating mass exodus. Ask a sperm whale, sea otter, or lynx if it is a component. By dividing and reducing the value of nature into parts and pieces, the UN Convention overlooks how animals interweave the components of the living world into the Blue Marble we ogle from space but ignore up close. Physicist Werner Heisenberg remarked, "What we observe is not nature in itself, but nature exposed to our method of questioning." If the loss of the majority of the living world is a response, what was the question? Heisenberg criticizes scientism as a method of knowing. Animal and plant communication, cognitive awareness, and intelligence are beyond our understanding. The untranslatable world exists regardless of belief or interpretation. Hannah Arendt wrote how the public square, the urban and rural spaces where people meet face-to-face, was crucial to human dignity and understanding. The same could be said of people and animals. Animals are justifiably secretive and vigilant. To restore ecosystems, animals need the ability to roam, move about, and meet each other. We need to meet them in ways that are safe, honoring, and kind.

Biodiversity is the constant interaction between creatures, large and small, the entire system of life and its interrelatedness, not a list of its "parts." The dance of carbon is the organizing, rearranging, pooping, munching, recycling, inhabiting, burrowing, nesting, pollinating, and sustaining of every ecosystem on the planet. Uncountable forms of life nurture habitats in grasslands, rivers, glens, wetlands, fertile soils, coral reefs, mangroves, and forests. Their home *is* our home. What we eat, see, smell, use, and depend upon is ultimately owed to animals—even unlikely items like dishwashers, eyeglasses, diapers, and the internet.

Plants metabolize energy from the sun. Animals metabolize energy from plants, eaten directly or by consuming plant-eaters. Energy is the currency of life, transmitted in the form of carbon into sugars, fats, and proteins. Ecosystems are interlocking elements of larger energetic systems. The interaction of animals, vegetation, and fungi is reciprocal. This exchange of energy, this core relationship, is the basis of life. We are entirely dependent upon and intertwined within those systems. The phenomenal diversity of species ensures that the energy from the sun is captured on every level, from a lichen to a lion. When you watch a video of marching leafcutter ants carrying pieces of foliage, imagine them carrying fifteen-hundred-pound bags of groceries from the forest store to their home. In this, they are like us, only stronger. We have towns and cities; they establish colonies holding ten thousand million (leafcutters are called town ants in Texas). We have farms; they cultivate crops of fungus growing on beds of chewed vegetation. We have vocations; they have complex divisions of labor. Trillions of animals engineer and construct the scaffolding of planetary life, yet we, another animal, are knocking them off, fishing them away, plowing them to death, clear-cutting them aside, and dehydrating them into oblivion. The casual reference to the "importance" of biodiversity is a monumental understatement.

The losses are not always apparent. The destruction of beavers in the United States desiccated hundreds of square miles of fens, bogs, and marshes. Wetlands support plants, crustaceans, frogs, wading birds, turtles, muskrats, beetles, mollusks, dragonflies, and hundreds of other species. Tallgrass prairies that covered one third of the United States are reduced to less than 4 percent of their range. The shifting baseline syndrome means people see gradual changes only over their lifetime. Unless it is a recent event, no one sees a missing wetland. The Earth we inhabit is accepted as normal because we lack a historical perspective.

Imagine that animals know more about the planet than human beings. How could they not? Is there a level of interspecies communication far beyond our understanding? When Monica Gagliano visited the reef on the last day of her experiment, the damselfish knew she was there to terminate their lives. This is an anecdote and has no scientific proof. The meaning was between the fish and Monica, the scientist. Science leads to extraordinary truths, but not the only truths. Most Indigenous cultures experience trees and animals, even water itself, as living beings, a community of which they are a member. European settlers treated that as superstition. Native cultures know it as reality.

In 1970, men corralled a pod of orcas into Penn Cove in Puget Sound using speedboats and spotter planes. They lobbed perchlorate seal bombs into the water, stunning the pod with two-hundred-decibel shock waves. Juveniles were separated from their mothers and pulled out of the water into nets one by one, uttering shrill, piercing cries. They had never experienced entrapment or feeling their body weight out of water. A mother and three juveniles died. The herders attempted to hide the deaths by slitting open the orcas' stomachs and filling them with rocks in hopes they would sink. They did not. It was a massacre. An orca pod is a family, not a group. One juvenile orca, known as Tokitae by the Lummi people, was sold and shipped to the Miami Seaquarium for $20,000. The twenty-foot-long orca was kept in a tank the size of a hotel swimming pool for fifty-three years with no protection from the sun. After decades of lawsuits and actions by tribal groups, the US government, and animal rights organizations, Tokitae was released in 2023. She died shortly thereafter, before she could be brought home. For the coastal-dwelling Lummi, orcas are sacred and are integral to their family as cousins, children, and parents. There was grief. Tokitae was cremated without their permission. In a private ceremony, her ashes were scattered into the Salish Sea by the Lummi. Orcas came to watch. Some of the Lummi believe Tokitae's mother was there. Known as Ocean Sun, she is nearly one hundred years old.

The word *biodiversity* is a bloodless term. Platitudes and jargon about nature mask atrocities. Honest language can be powerful, poignant, and uplifting in the hands of masters like Robin Wall Kimmerer, Ralph Waldo Emerson, Carl Safina, Merlin Sheldrake, Mary Oliver, Barry Lopez, Linda Hogan, David James Duncan, and others. The word *wild* conjures diverse reactions. Some think wild means crazy or uncontrolled. Not so. We are wild, each of us, as in original, innate, authentic, instinctive, and deep-rooted. The overwhelming array of industrial forces lined up against the living world—the pipelines, mines, poisons, pharmaceuticals, agriculture, trawlers, plastic, and banks—favor uniformity, sameness, repetition, control, hierarchy, force, violence, and even oppression. That is not wild. It is death.

Wild is exquisite. Wild is a life that benefits other beings. Wild is never crazy. Crazy is double-glazing the planet with the carboniferous era, killing the oceans with carbonic acid, sterilizing our soils, and then genetically manipulating seeds to fix our stupidity. The unraveling of our home planet is a mental disorder in which thoughts and feelings are so impaired they have no relationship to external reality. Wild is the opposite. It is when our thoughts, feelings, and actions are exquisitely sensitive to our relationship with the living world and each other. Wild is why we dance, write, sing, protest, play sports, climb mountains, stand on our heads, and become slam poets. It is the feeling of our feet as the earth, our eyes as the sky, our hearts like headwaters, our breath the atmosphere. We are denizens. Let's come home.

There are tens of thousands of environmental and social justice organizations. They *are* wild, social organisms comprised of diverse, adaptive, compassionate, fierce, local people. See them as one might encounter a swarm of eels, a skein of geese, a scold of jays, or a flutter of monarchs. The kingdom of life is reasserting itself within human awareness. As anxiety and panic course through generations, there is an opening. The trauma of being in the world today makes sense. It is the appropriate response. It can lead to closure or opening. Alexa Firmenich believes our sadness and distress are a homecoming to our naked selves, a place where we see how beautiful the world truly is. Báyò Akómoláfé turned the table in a commencement speech: "We must allow that our lives are not durationally competent enough to hold all the questions we could possibly explore, for lives and deaths are not matters of duration alone. Listen to your failures, don't cover the cracks up, go deep in there. Whatever you do, don't try to make the world a better place; instead, consider that the world might be trying to make you a better place."

FIFTEEN

Conscious

*Sit, be still, and listen, for you are drunk, and
we are at the edge of the roof.*

JALĀL AL-DĪN AL-RŪMĪ

The third line of Walt Whitman's poem "Song of Myself,"
reads, *"For every atom belonging to me as good belongs to you."*
The line echoes traditions of native cultures whose bonds to
land, plants, and animals are inseparable. The whole of life—
birds, fruits, snakes, trees, moths, grasses, and leopards—are
members of their communities. As has been told to me, knowledge of the community is the ongoing expression of mentorship—
shared understandings offered by parents, elders, siblings, and
family that emerged from being on the land. It is an experience, a continuity, a flow, not a teaching, ecological wisdom
passed down and modified over generations. Today, civilization reflects almost complete lack of awareness. The dominant
world mistakes community for commodity, and exploits even

itself. I don't know a word that describes humanity's yawning disconnection from the living world. The climate movement is alive and growing, populated by sincere people, but it cannot succeed unless we see the planet as a living entity, one and the same as earthworms, lichen, and lemurs. Life must be at the center of all we do or we will not live here much longer. "Every atom belonging to me as good belongs to you."

I once walked on the unceded land of the Wampanoag near Cape Cod, following a winding, unused dirt road. As I rounded a bend, seven animals faced each other in a circle. It appeared to be a meeting, like a council. I froze. So did they, for a moment. There was a hognose snake, a box turtle, a possum, a rabbit, a white-footed mouse, a meadow vole, and two bob-white quail. Those that could bolted or slithered into the hazelnut and sea oats. The turtle lumbered away, its leathery tail leaving a wavy line in the thick dust. I couldn't believe what I had seen. I checked their tracks. They were there. There is no explanation. The event has stayed with me. People who spend time in nature have experiences for which there is no logical explanation. The natural world is an antidote to the disarray and madness that infest the communications that surround us, a sanctuary of truth.

I am not qualified to explain the wisdom, mores, and customs of the myriad cultures that populated the Earth before the colonial holocaust. I am a European mutt with genes that sprawl across the map, schooled in traditional Western beliefs. No ancestral wisdom was passed on to me. However, my grandparents were of Scottish, Swedish, Cornish, and Alsatian descent, decent people, and quite pragmatic. They were primarily farmers. I was taught to distinguish what works and what doesn't, to be practical, which is why I defer to cultures that have inhabited lands far longer than my known ancestors. Hindou Omarou Ibrahim, a Wodaabe pastoralist from Chad, was asked if her culture planned seven generations ahead in their decision-making. The idea and term "seven generations" originated with the Great Law of Peace of the Haudenosaunee Confederacy in the late sixteenth century, where five nations joined to harmonize their political, ceremonial, and social foundations. When a decision is agreed upon by the fifty clan chiefs of the Onondaga, Oneida, Cayuga, Seneca, and Mohawk (Kanienkehaka) nations, it is referred back to the Onondaga to determine if the decision would be in accordance with the Great Law. The Great Law determines whether an action would benefit or harm the welfare of the people seven generations into the future. Would an action reduce and endanger resources, or would it protect and amplify them? Bear in mind, for Indigenous people, one generation comprised seventy years. Pragmatism means dealing with things realistically, the future well-being of the nation. There could be no more practical covenant than the Great Law.

When Ibrahim responded with a yes, that the Wodaabe recognized the principle of the Great Law, she added an important qualifier. "We can do this because we remember seven generations into the past." She spoke not as an individual but as someone who has an enduring relationship with the past. How does she recall decisions and events that far back? She doesn't.

Her community does. Lore, teachings, and information are retained collectively. Memory is encoded in stories, songs, and art that go back centuries. Recall the Diné's seven hundred accurate descriptions of insects with no written language. For most Westerners there was scant, if any, ecological wisdom to be remembered. The proper relationship to living beings was not taught. Aside from the Haudenosaunee Confederacy, which is the oldest democracy in the world, there is no major Western institution that makes the future of all living beings its governing principle.

Our genus, *Homo sapiens*, became dominant because we weren't initially. We prevailed because we built networks and problem-solved collectively. *Sapiens* means astute or wise, an appendage bestowed by taxonomist Carl Linnaeus. It is difficult to believe the modifier is merited, given where we are today. The world, especially the United States, has seen a steep decline in comity, tolerance, and mutual understanding. Jonathan Haidt writes, "We are disoriented, unable to speak the same language or recognize the same truth. We are cut off from one another and the past . . . the scattering of people who had been a community." Can Humpty-Dumpty societies be put back together again? Haidt's comments presume there was a community. Colonist nations have been exterminating communities for five hundred years. No distinction was made between beavers, the Taino, right whales, or the Cherokee. Consequences from past actions surround us—economic warfare, standing armies, simmering oceans, pervasive anxiety, flattening storms, declining mental health, and tyrannous heat—symptoms of the loss of community. Dime-store beliefs drop when mocked by reality. I do not recall climate deniers, volunteers, or first responders arguing about political or religious beliefs during or after a significant storm, fire, or disaster. What brings people together is the desire to share food, water, shelter, warmth, kindness, and community. Our social and environmental ecologies are inextricable. The failures of politics and society are an invitation to reimagine each other within the living world.

In the United States, a preeminent example of an Indigenous community is the Black church, a devout culture that emerged from the diaspora of enslavement, a culture that has transformed the world through leadership, music, literature, dance, sports, and art. In 1965, I worked in Selma, Alabama, for the March to Montgomery led by the Southern Christian Leadership Conference. I was nineteen, a white volunteer, and my role was minor. Its impact on me was indelible. Having traveled throughout Europe for a year when I was seventeen, I encountered in Alabama a culture vastly different from anything I had seen in the twelve countries I visited in Europe. The Brown Chapel African Methodist Episcopal Church, founded in 1866, was the heart and center of the march, a centuries-old community built as a chalice of goodness by people who had been persecuted and defiled by slavery, racism, and violence. Night and day until the march, there were sermons, testimonies, gospel, and singing.

You felt resolve in the church, a pulsing, unwavering determination to overthrow racism in all its manifestations. The unwavering tenacity was amplified by the beating and murder of unarmed Baptist deacon Jimmie Lee Jackson by Alabama state troopers in a nearby café four weeks earlier. It was another lynching. From the balcony pews, I watched parishioners transfixed by pastors like Martin Luther King Jr., James Bevel, and Andrew Young. The rhetoric was extemporaneous, eloquent, and compassionate: "We must not be bitter, and we must not harbor ideas of retaliating with violence. We must not lose faith in our white brothers" was part of King's startling requiem for Jimmie Lee Jackson. I had never witnessed unshakeable faith—the amens, head-nodding, hallelujahs, and upheld waving hands. There was sorrow and grief. There was joy and absolution. There was mercy. There was grit. And there was singing, the kind of church choir that produced Aretha Franklin, Whitney Houston, and Jennifer Hudson. It was a community I had never seen or imagined. I was a white person in the church. They did not know who I was, yet I was welcomed with kindness. It was quite different from the funereal masses I tended to as an altar boy at St. Mary's Catholic Church in Oakdale, California. Parishioners in the Brown Chapel had been mocked, demeaned, and denigrated; I believe the majority of men in the church had been abused and humiliated in front of their wives and children by white men of Selma at some point in their lives. There were mothers there who had gone to hospitals to see their disfigured sons who had done nothing wrong but exist in dark skin. However, I did not experience any expression of blame, revenge, or self-pity. What I saw was dignity.

I think of the Brown Chapel because it asks questions: What does it mean to be a human being at a time when the fabric of life is being shredded? Who are we, and what are we capable of? Will we do something? Will we hope someone else does something? I believe most people do not understand the planetary and social perils we face. Even if they knew, they might not understand the causes. Can communities of action that are compassionate, effective, and brilliant emerge in the growing chaos? Einstein famously said that the most essential question for humanity is whether the universe is friendly or not.

Where do we place the fragile, fleeting gifts of our life? With those who speak the truth, who create authentic relationships with society and other living beings, an assemblage of humanity that is willing to stand up to the raw, ceaseless insults that come from the guns, checkbooks, and policies of corporations and governments. Those who do not feel grief at the state of the world are bereft, for grief is a measure of one's love. The late poet Mary Oliver concedes there can be a life without love, but "it is not worth a bent penny, or a scuffed shoe. It is not worth the body of a dead dog nine days unburied."

How can the truths of the living world be shared if most people are urbanized and do not experience the world around them? As with doctors and healers, there are people who have devoted their lives to the knowledge and understanding we lack. They are more common than one might think. They are naturalists, scientists, Latinos, birders, farmers, African Americans, and Indigenous citizens around the world. They want to share and teach. We are not alone in this. If you turn to the living world, it will turn to you.

There are wisdom keepers among us whose cultures are

based on reciprocity. Where reciprocity prevails, everyone benefits. When reciprocity is absent, injustice prevails. Thousands of native communities that were treated as the enemy, who endured measureless brutality and atrocities, and whose people were removed from lands where they had lived continuously for more than fifty thousand years are now reclaiming their rightful place as progenitors of respectful and restorative wisdom. It is instructive to talk to people who endured centuries of apocalyptic repression at the hands of settlers and colonizers. Native cultures were annihilated en masse; their land was stolen, their children taken, and their language banished. In Botswana until 1963, one could purchase a hunting license to kill the San people. Yet they survived because reciprocity was woven into everyday activities. These included ritual, self-reflection, nourishment, humor, honesty, fairness, hard work, respect for elders, and love for children. The San endured because of the actions taken by those who came before them. They are here because they embody ways of knowing rooted in shared values. Indigenous ecological knowledge is far more relevant than the proceedings of the annual World Economic Forum or the deliberations of the UN. To put it plainly, the teachers we need are here, not in Davos or New York City.

The cascade of troubling information about the future is staggering and dispiriting. Great damage has been done to the whole of the living world. Chief Oren Lyons of the Onondaga Nation described a prophecy that foretold a time when the Earth became biologically depleted and the purpose of humanity was lost. There would be two signs. One, the wind would howl and blow as never before. The envelope of the earth would be ripped and torn by the activity of humankind. The second sign would be children. They would be abandoned and ignored. Is this any longer a prophecy?

Most of us turn down the dial to function. "Consciousness, the great poem of matter, seems so unlikely, so impossible, and yet here we are with our loneliness and our giant dreams," writes Diane Ackerman. People sense something momentous is happening. Barry Lopez wrote, "We feel ourselves on the verge of something vague but extraordinary. Something big is in the wind, and we feel it. . . . We know that if we mean to make this a true home, we have a monumental adjustment to make. . . . We must turn to each other and sense this is possible."

Lopez counseled, "If you are afraid of what might happen in the future, find a person you respect who does not act out of fear." If you feel overwhelmed, read the biography of Sojourner Truth or Cesar Chavez. If you think being kind, respectful, and polite is ineffective, listen to Jane Goodall and Robin Wall Kimmerer. If you feel ineffective, mentor a child, heal a wounded animal, feed the hungry. If you are weary of chasing hope, read *Original Instructions*, written and edited by Melissa Nelson, a member of the Turtle Mountain Band of the Chippewa. To stop the mind from caving in on itself, go outside. Replace digitized awareness with direct experience. Touch things. Mend and revive a verge, some sullied land, a habitat, your backyard, a relationship. While we storm the Bastille of corporate ignorance and political corruption, introduce native plants to your environment that provide food and sanctuary for pollinators and birds. Learn their names and stories. Wendell Berry counsels us to be joyous though we know all the facts. Although humanity faces what appears to be an insurmountable endgame brought about by ignorance, aggression, and greed, we also live in the most brilliant period in human history. Renewal results from disturbance. There are astonishing breakthroughs arising from breakdowns, ways of seeing and acting in the world that bring swaths of humanity back to life. The gift that rests at the heart of the myriad crises is newfound purpose. Everyone craves a life of meaning. Regenerating the world is the journey to possibility. Vistas open. The diverse voices, social organisms, and entities now emerging worldwide are rehearsing the future. As I wrote this sentence, a swallowtail butterfly flitted through the window, its wings gently fanning the air above the keyboard, and then it flew on.

Change and wonder, doubt and fear, walk hand in hand. This is the nameless era. It was predicted, but the common fate of prophecy is to be ignored. The juggernaut institutions laying waste to sea, land, and people cannot endure. Top-down corporate solutions for saving life on Earth are well intended but will fail because nature is not top-down.

A beginning is near, a threshold, and so too is an end. Without fail, meaningful change begins with one person, one idea, one aspiration, one audacious dream. Uniqueness is your birthright, it is the seed of community. Plant it and see what happens. Pessimism and gloom are cobwebs; brush them aside. We seek a rapprochement with Mother Earth, what Stephan Harding calls "the vast and mysterious primordial intelligence that steadily gives birth to all that exists—that sustains all that is." We eat, drink, love, and breathe because of this mantle of life. Do we cherish it or lose it? You can't be both cautious and courageous, we must choose. Focus on what is in front of you. Give yourself permission to fail. Leave room for foibles, humor, and giggles. Find restorative movements you can sing and dance to, lest creation "plays to an empty house."

Beliefs do not change our actions; actions change our beliefs. Complex realities begin as simple acts—enchantment, humility, respect, imagination, and constant gratitude—which offer wider apertures to the living world. Monica Gagliano suggests that we stop playing God and instead play midwife. We can't save the planet; it will save itself. It is innately regenerative. We are invited to honor the world that is right in front of us. The living world is your best friend.

Where you are is where you are most effective. The power to act does not lie elsewhere. Fundamental human rights and needs must be met. Everyone on Earth comes first; there is no second. Revive, honor, and nourish the wild and bountiful lives that forever astonish us with their splendor and grace. The movement to restore life on Earth is not a repair job. It is transformative, an entirely new experience of self, the visceral awareness that our life is coincident with every being on the planet. Our intention and reward are the same: to experience and express the irrevocable connection to all beings. It is our only way forward.

Acknowledgments

Creating a book is a process of gathering, listening, questioning, reading, and observing. A nucleus forms, a tentative title perhaps, and what was a notion becomes a seed that starts to delineate and emerge into a structure with interconnected flows. The bones of the book may fall in place but success is never included. A book can take a year, or a decade. My books begin with curiosity, a desire to discover. They do not result from expertise or extensive prior knowledge. *Carbon* is not a book about what I know. I pray it is a book about what is known. For that, relationships play a crucial role. This acknowledgment is an expression of gratitude to the writers, friends, dreamers, poets, teachers, and scholars who have graced my life and fed my mind.

Foremost are Joe Spieler, my lifelong agent—a trusting, patient, and wise soul—who never pushed a timetable, and my droll, comprehensively literate editor, Rick Kot, who never doubted my resolve. *Carbon* was first signed and contracted by Rick more than ten years ago. It was delayed by my wanting to author two other books first, *Drawdown* and *Regeneration*. During that decade, Joe and Rick were calm and accommodating. I am pleased to honor their faith. Both men are sanctuaries of scholarship and kindness. And a big bow of gratitude to Allison Lorentzen and her team at

Viking, who seamlessly took over editorial responsibilities as Rick Kot was retiring. I felt sustained in every way by their competence and belief in the book.

To my family members, Palo Hawken, Anastasia Hawken, and especially those who were directly engaged: my beloved wife Jasmine, daughter Iona, and son Jonathan. I was blessed by their unswerving support. I sometimes falter and question the merit of my writing. They never did. I was inspired and moved by some extraordinary friends: David James Duncan teaches me how to write every time I hear him speak or read his work. From David I learned to focus on that which is worth knowing. On the other side of "knowledge," he reminds me to never forget, in the words of poet Khwaja Mir Dard, that there is no scarcity of angels amongst us. I cannot sufficiently emphasize the influence of Barry Lopez and express how sorely missed and treasured he remains for so many who work to restore and honor the natural world. His importance will forever grow. I am honored by my Turtle Band Chippewa neighbor Melissa Nelson, whose sage advice exudes the emergent vitality and wisdom of First Nations. I am fed by my dear friend Anthony James, whose podcasts joyously walk a path through the Earth's dilemmas. I am uplifted by Alexa Firmenich, who merges the outer and inner landscapes of our troubled civilization so eloquently. I am amazed by Javier Pena, my Spanish brother, who tirelessly animates millions of people into action by his deep knowledge, caring determination, and guileless charm. And I have unending respect for Chhaya Bhanti, an astonishingly effective activist in India, who is transforming the ecology of energy, water, and farming in northern India.

The most outstanding person I know in the field of regeneration is Damon Gameau in Australia. A superb and dedicated filmmaker, he has a remarkable and unmatched capacity to build community, knowledge, and action by storytelling. I have been sustained by the decades-long influence of longtime friend Van Jones, whose depth of wisdom never ceases to grow in stature and breadth. I am joyously uplifted by AY Young, the hippest songster of the regenerative movement, who gives more to his audience in one performance than can be imagined. I look to the dearly beloved Jane Goodall, the wise Earth elder, who has given the whole of her life to the well-being of others. She is the beacon on the hill where all human compasses should point. Báyò Akómoláfé's endlessly imaginative mind continues to animate and surprise, opening and revealing the narrows of my thinking. He alters every aspect and concept of modernity including the English language itself in ways that leave one startled and spellbound.

Two people in particular stood alongside me, thick and thin on this journey. I met Julia Jackson when she walked into the green-room years ago after a speech holding one of my books, yellow-lined, dog-eared, its binding failing. Since then, she became an unrelenting supporter of my work. She also transformed what other cultures might call a shaman—a calm, quiet teacher of nonduality, embracing and ministering to the earthly and social travails the world faces. Katie Gray has devoted her life to helping others rediscover that the love we crave resides profusely in our hearts. At a time when many feel profoundly disconnected from the social and natural world, her presence teaches us how to come home.

There are writers I may never meet that catalyze my neuronal pathways with their profound intelligence and depth. They include Merlin Sheldrake, Andri Snær Magnason, Melanie Challenger, Amitav Ghosh, Monica Gagliano, Robin Wall Kimmerer, Peter McCoy, Dara McAnulty, Camilla Pang, Ed Yong, Eric Roston, Carl Zimmer, the late Karen Bakker, Fred Provenza, Zoë Schlanger, Stefano Mancuso, and Stephen Buchmann. Michael Pollan and David Montgomery, whom I do know, belong with aforementioned writers.

My deepest gratitude to Isabella Tree and Charlie Burrell for hosting me at the Knepp Estate. Their work on rewilding is nothing short of spectacular. We will look back one day and see that their brilliant, unassuming discoveries are commensurate to that of Alfred Russel Wallace and Jane Goodall. I honor Crown Princess Victoria of Sweden for her friendship and unflagging dedication to the environment and human well-being. The cover image of this book is a gift from an extraordinary photographer and friend in Chile, Chris Jordan. I have a few hundred of his images on my hard drive and each one invites and deserves a book and a cover. They are singular and I am deeply grateful. Toby Kiers, a wizard of the fungal world inspired an entire chapter. Her effervescence and joy can make one a fungaphile forever.

Throughout, I was supported by many friends including Haley Melin, Aileen Getty, Tad Buchanan, Liangbing Hu, Alex Lau, Erik Snyder, Josh Felser, Liesl Copland, Adam Parr, Rob Cameron, Peter Coyote, Bill McGlashan, Zana Briski, Alec Webb, Michael Stusser, Cynthia Hardy, John Hardy, Brandee Alessandra, Martin Goebel, Damien Sabella, Per Espen Stoknes, Megan Camp, Justin Winters, Michelle Best, Jib Ellison, Elson Haas, Cecily Mak, Federico Mennella, Charles Massy, Lilian Riesenfeld, Dave Chapman, Mark Hyman, Janet Mumford, Soren Gordhamer, Kathryn Marshall, Rachel Vestergard, Maisa Arias, and many more. Forgive me if your name is not here.

A bow to Jack Kornfield, Tara Brach, and Jon Kabat-Zinn, who constantly remind me that while personal, social, and global dilemmas arise because human desires are endless, the human capacity for compassion, kindness, and selflessness is also endless.

I am forever beholden to Roz Zander, whose contribution, wise counsel, and graciousness nurtured me for decades, right up until her untimely death one year ago. I will never forget her.

Notes

ONE: CARBON

1 **"There are things we must do"**: Báyò Akómoláfé, "Welcome, Traveller: Foreword," Báyò Akómoláfé, bayoakomolafe.net.

2 **There are women and men**: Camilla Pang, *Explaining Humans: What Science Can Teach Us about Life, Love and Relationships* (London: Penguin Books, 2021).

2 **I turned to voices**: Báyò Akómoláfé and Indy Johar, "The Edges in the Middle, III: Akómoláfé, Báyò and Johar Indy," in *For the Wild*, May 24, 2023, produced by For the Wild, podcast, MP3 audio, 58:51, forthewild.world/listen/the-edges-in-the-middle-bayo-and-indy.

4 **the value of a blue whale**: Ralph Chami et al., "Nature's Solution to Climate Change," *Finance & Development* 56, no. 4 (December 2019): 34–38, imf.org/external/pubs/ft/fandd/2019/12/pdf/natures-solution-to-climate-change-chami.pdf.

4 **commerce is eliminating life**: "Destroying the Earth to pay dividends" is a rewording of a phrase from Gabor Maté's talks.

4 **"we need a change of worldview"**: Robin Wall Kimmerer, "The Turtle Mothers Have Come Ashore to Ask about an Unpaid Debt," *The New York Times*, September 22, 2023, nytimes.com/2023/09/22/opinion/climate-change-turtles-refugees.html.

5 **The task of modernity**: Akómoláfé and Johar, "The Edges in the Middle, III."

5 **The world economy:** Eric Roston, "Corporate Net-Zero Goals Don't Add Up to a Net-Zero Planet," *Bloomberg*, June 27, 2022, bloomberg.com/news/articles/2022-06-27/companies-net-zero -emissions-goals-don-t-add-up.

5 **"We are trying":** Melanie Challenger, *How to Be Animal: A New History of What It Means to Be Human* (New York: Penguin Books, 2021), 211.

5 **In all of Earth's:** Judith Schwartz, "Doing the Impossible," in *The RegenNarration*, episode 175, August 1, 2023, produced by The Regen-Narration, podcast, MP3 audio, 1:22:44, regennarration.com/episodes /175-judith-schwartz.

6 **"Our cities and industries":** Challenger, *How to Be Animal*, 2.

6 **Replacing fossil fuels:** Julia Janicki et al., "The Collapse of Insects," Reuters, December 6, 2022, reuters.com/graphics/GLOBAL -ENVIRONMENT/INSECT-APOCALYPSE/egpbykdxjvq. "Carbon tunnel vision" was coined by Jan Konietzko in "Moving Beyond Carbon Tunnel Vision with a Sustainability Data Strategy," Cognizant, February 8, 2022, digitally.cognizant.com/content/digi tally-cognizant/us/en/blogs/moving-beyond-carbon-tunnel-vision -with-a-sustainability-data-strategy-codex7121.html.

6 **the dance of carbon:** Eric Roston, *The Carbon Age: How Life's Core Element Has Become Civilization's Greatest Threat* (New York: Walker, 2008). This is an older book but carbon never dates. Without question the best book about the subject by a world-class journalist. The book has a chapter entitled "Dancers and the Dance: The Origins of Life."

7 **cellular community containing:** Ian A. Hatton et al., "The Human Cell Count and Size Distribution," *Proceedings of the National Academy of Sciences* 120, no. 39 (September 18, 2023): e2303077120, doi.org /10.1073/pnas.2303077120.

8 **Western science became:** Monica Gagliano, *Thus Spoke the Plant: A Remarkable Journey of Groundbreaking Scientific Discoveries & Personal Encounters with Plants* (Berkeley, CA: North Atlantic Books, 2018), 87; Ray Archuleta, "Plant and Soil Are One," presented at Natural Resources Conservation Service Training, Ames, Iowa, Spring 2014, video, 1:30:16, youtube.com/watch?v=FQEKlm4DOdw.

9 **"May this decade bring":** Báyò Akómoláfé, "Welcome, Traveller."

TWO: ELEMENTS

11 **Carbon is the most mysterious element:** Eric Roston, *The Carbon Age: How Life's Core Element Has Become Civilization's Greatest Threat* (New York: Walker, 2008), 28.

12 **When we digest:** Annie Dillard, *An American Childhood* (London: Canongate, 2016), 28. "Skin was earth; it was soil. I could see, even on my own skin, the joined trapezoids of dust specks God had wetted and stuck with his spit the morning he made Adam from dirt."

12 **No matter what we believe:** Matthew J. Shribman, "The Biggest Communication Failure in History," *Matthew Shribman (CliMatt)*, September 29, 2023, matthewshribman.substack.com/p/the-biggest -communication-failure.

13 **The prediction that increased atmospheric:** Clive Thompson, "How 19th Century Scientists Predicted Global Warming," *JSTOR Daily*, December 17, 2019, daily.jstor.org/how-19th-century-scientists-pre dicted-global-warming.

14 **Although it makes up:** Roston, *The Carbon Age*, 26.

16 **As I write this:** Caroline Hickman et al., "Climate Anxiety in Children and Young People and Their Beliefs about Government Responses to Climate Change: A Global Survey," *The Lancet Planetary Health* 5, no. 12 (December 2021): e863–e873, doi.org/10.1016/S2542-5196(21) 00278-3.

16 **In 2021, an international survey:** Hannah Ritchie, "Stop Telling Kids They'll Die from Climate Change," *Wired*, November 1, 2021, wired.com/story/stop-telling-kids-theyll-die-from-climate-change.

16 **The climate message is toothless:** Shribman, "The Biggest Communication Failure in History."

16 **Among the billions of fish:** "Migration in the Ocean Twilight Zone," Woods Hole Oceanographic Institution, twilightzone.whoi .edu/explore-the-otz/migration.

16 **In the ocean:** Allen Collins, "What Is Vertical Migration of Zooplankton and Why Does It Matter?," National Oceanic and Atmospheric Administration, October 28, 2021, oceanexplorer.noaa.gov /facts/vertical-migration.html.

17 **The oceanic surface layers:** Shribman, "The Biggest Communication Failure in History."

THREE: FIRMAMENT

21 **Over billions of years:** Robert M. Hazen, *Symphony in C: Carbon and the Evolution of (Almost) Everything* (New York: HarperCollins, 2019), 19.

22 **"a sheaf of empty space":** Melanie Challenger, *How to Be Animal: A New History of What It Means to Be Human* (New York: Penguin Books, 2021), 200.

22 **In 2023, images from:** Daniel Clery, "Earliest Galaxies Found by JWST Confound Theory," *Science* 379, no. 6639 (2023): 1280–81, doi:10.1126/science.adi0089.

24 **Looking back at that time:** Natalie Wolchover, "A Primordial Nucleus behind the Elements of Life," *Quanta Magazine*, December 4, 2012, quantamagazine.org/the-physics-behind-the-elements-of-life -20121204.

25 **Once it was shown:** Wolchover, "A Primordial Nucleus behind the Elements of Life."

26 **The mantle of Earth:** "Scientists Catalogue Earth's Total Carbon Store," BBC, October 1, 2019, bbc.com/news/av/science-environment -49899042.

27 **Today, when physicists:** Fred Hoyle, *The Nature of the Universe* (New York: Harper & Brothers, 1950); Eric Roston, *The Carbon Age: How Life's Core Element Has Become Civilization's Greatest Threat* (New York: Walker, 2008), 6; Wolchover, "A Primordial Nucleus behind the Elements of Life."

27 **If any one of these properties:** Stephen C. Meyer, *Return of the God Hypothesis: Three Scientific Discoveries That Reveal the Mind Behind the Universe* (New York: HarperCollins, 2021), 204, Kindle.

FOUR: CELL MATES

30 **In school, we were taught:** Merlin Sheldrake, *Entangled Life: How Fungi Make Our Worlds, Change Our Minds & Shape Our Futures* (New York: Random House, 2020), 210.

32 **a term coined by:** Laura Poppick, "Let Us Now Praise the Invention of the Microscope," *Smithsonian*, March 30, 2017, smithsonianmag.com /science-nature/what-we-owe-to-the-invention-microscope-180962725.

32 **The microscope master:** Karen Bakker, *The Sounds of Life: How Digital Technology Is Bringing Us Closer to the Worlds of Animals and Plants* (Princeton, NJ: Princeton University Press, 2022), 376, Kindle.

33 **"If viruses are lifeless":** Carl Zimmer, *Life's Edge: The Search for What It Means to Be Alive* (New York: Dutton, 2021).

35 **Physical sciences divide:** Melanie Challenger, *How to Be Animal: A New History of What It Means to Be Human* (New York: Penguin Books, 2021), 217.

36 **Mars had been the object:** Erik Washam, "Cosmic Errors: Martians Build Canals," *Smithsonian*, December 2010, archive.today /20120912195828/http:/www.smithsonianmag.com/science-nature /Cosmic-Errors.html.

36 **The term was mistranslated:** David W. Dunlap, "Life on Mars? You Read It Here First," *The New York Times*, October 1, 2015, nytimes .com/2015/09/30/insider/life-on-mars-you-read-it-here-first.html.

36 **In 1895, Percival Lowell:** Kat Eschner, "The Bizarre Beliefs of Astronomer Percival Lowell," *Smithsonian*, March 13, 2017, smithsonianmag .com/smart-news/bizarre-beliefs-astronomer-percival-lowell -180962432.

38 **"The Earth is essentially":** Eric Roston, *The Carbon Age: How Life's Core Element Has Become Civilization's Greatest Threat* (New York: Walker, 2008), 28.

40 **We need to be quiet:** Avalon C. S. Owens and Sara M. Lewis, "Artificial Light Impacts the Mate Success of Female Fireflies," *Royal Society Open Science* 9, no. 8 (August 2022): 220468, http://doi.org/10 .1098/rsos.220468.

40 **Sound and light bewilder:** Annika K. Jägerbrand and Kamiel Spoelstra, "Effects of Anthropogenic Light on Species and Ecosystems," *Science* 380, no. 6650 (June 15, 2023): 1125–30, doi:10.1126/science.adg3173.

40 **Two thirds of invertebrates:** Johan Eklöf, *The Darkness Manifesto: On Light Pollution, Night Ecology, and the Ancient Rhythms That Sustain Life* (New York: Scribner, 2024), 24, Kindle.

40 **The terrorist assault:** Ed Yong, *An Immense World: How Animal Senses Reveal the Hidden Realms around Us* (New York: Random House, 2022), 340, Kindle.

41 **The sensitivity shown:** Annie Novak, "The 9/11 Tribute in Light Is Helping Us Learn about Bird Migration," All About Birds, Cornell Lab of Ornithology, August 30, 2018, allaboutbirds.org/news/9-11 -tribute-in-light-birds-night-migration.

44 **When anthrophony converges:** Karen Bakker, *The Sounds of Life: How Digital Technology Is Bringing Us Closer to the Worlds of Animals and Plants* (Princeton, NJ: Princeton University Press, 2022), 14, Kindle.

44 **Conversely, Krause recently returned:** Phoebe Weston, "No Birdsong, No Water in the Creek, No Beating Wings: How a Haven for Nature Fell Silent," *The Guardian*, April 16, 2024.

44 **Nature has listened to itself:** Bernie Krause, *Voices of the Wild: Animal Songs, Human Din, and the Call to Save Natural Soundscapes* (New Haven, CT: Yale University Press, 2015), 29–30, Kindle; Krause, *Voices of the Wild*, 25–26, Kindle; Bakker, *The Sounds of Life*, 307, Kindle.

44 **based on those sounds:** Bernie Krause and Almo Farina, "Using Ecoacoustic Methods to Survey the Impacts of Climate Change on Biodiversity," *Biological Conservation* 195 (2016): 245–54, doi.org/10 .1016/j.biocon.2016.01.013.

44 **evolving language beyond our understanding:** Bernie Krause, "The Niche Hypothesis: New Thoughts on Creature Vocalizations and the Relationship Between Natural Sound and Music," *WFAE Newsletter*, June 1993.

44 **"a great silence":** Krause, *Voices of the Wild*, 11, Kindle.

45 **Although biology may not:** Nathan Robinson, "A Brain in Each Leg?," in *Conspiracy of Goodness*, episode 120, February 14, 2023, produced by Goodness Exchange, podcast, MP3 audio, 1:02:09, goodness -exchange.com/podcast/nathan-robinson-follow-curiosity.

FIVE: EATING STARLIGHT

46 **When we sip, bite, and chew:** Isaac O. Perez et al., "Speed and Accuracy of Taste Identification and Palatability: Impact of Learning,

Reward Expectancy, and Consummatory Licking," *American Journal of Physiology* 305, no. 3 (August 2013): R252—R270, doi.org/10.1152/ajpregu.00492.2012.

47 **For two million years:** "Indigenous Peoples: Respect NOT Dehumanization," United Nations, un.org/en/fight-racism/vulnerable-groups/indigenous-peoples.

47 **He landed in San Salvador:** Barry Lopez, *The Rediscovery of North America* (New York: Vintage, 1992).

47 **Europeans had experienced:** "List of Famines," Wikipedia, last edited on March 4, 2024, en.wikipedia.org/wiki/List_of_famines.

48 **three hundred thousand edible plants:** John Warren, *The Nature of Crops: How We Came to Eat the Plants We Do* (Wallingford, UK: CABI, 2015).

49 **Human bodies do not necessarily want:** "What Is Happening to Agrobiodiversity?," Food and Agriculture Organization, fao.org/3/y5609e/y5609e02.html.

49 **began a well-known study:** Stephen Strauss, "Clara M. Davis and the Wisdom of Letting Children Choose Their Own Diets," *Canadian Medical Association Journal* 175, no. 10 (November 7, 2006): 1199–201, doi:10.1503/cmaj.060990.

49 **"The nurses' orders":** Clara Davis, "Results of the Self-Selection of Diets by Young Children," *Canadian Medical Association Journal* 41, no. 3 (September 1939): 257–61, ncbi.nlm.nih.gov/pmc/articles/PMC537465/pdf/canmedaj00208-0035.pdf.

50 **The children's attending pediatrician:** Benjamin Scheindlin, "'Take One More Bite for Me': Clara Davis and the Feeding of Young Children," *Gastronomica* 5, no. 1 (February 2005): 65–69, doi:10.1525/gfc.2005.5.1.65.

50 **Fred Provenza wonders why:** Fred Provenza, *Nourishment: What Animals Can Teach Us about Rediscovering Our Nutritional Wisdom* (White River Junction, VT: Chelsea Green Publishing, 2018), 20, Kindle.

50 **Rats with laboratory-induced diabetes:** Kerstin Rohde, Imke Schamarek, and Mattias Blüher, "Consequences of Obesity on the Sense of Taste: Taste Buds as Treatment Targets?," *Diabetes & Metabolism Journal* 44, no. 4 (2020): 509–28, doi:10.4093/dmj.2020.0058.

50 **Today, sugary, ultra-processed food:** Tobi Thomas, "More Than a Billion People Worldwide Are Obese, Research Finds," *The Guardian*, February 29, 2004, theguardian.com/society/2024/feb/29/more-than-a-billion-people-worldwide-are-obese-research-finds.

50 **"sensory" chemists know what happens:** Nell Boeschenstein, "How the Food Industry Manipulates Taste Buds with 'Salt Sugar Fat,'" NPR, February 26, 2103, npr.org/sections/thesalt/2013/02/26/172969363/how-the-food-industry-manipulates-taste-buds-with-salt-sugar-fat.

51 **Ultra-processed foods are designed:** Lelia Green, "No Taste for Health: How Tastes Are Being Manipulated to Favour Foods That Are Not Conducive to Health and Wellbeing," *M/C Journal* 17, no. 1 (2014), doi.org/10.5204/mcj.785.

51 **ultra-processed food is directly linked:** Huiping Li et al., "Association of Ultraprocessed Food Consumption with Risk of Dementia: A Prospective Cohort Study," *Neurology* 99, no. 10 (September 6, 2022): e1056–e1066, doi:10.1212/WNL.0000000000200871.

52 **industry has flipped the script:** Chris van Tulleken, *Ultra-Processed People: Why We Can't Stop Eating Food That Isn't Food* (New York: W. W. Norton, 2023), 5–6, Kindle.

53 **Human beings can detect:** Caroline Bushdid et al., "Humans Can Discriminate More Than 1 Trillion Olfactory Stimuli," *Science* 343, no. 6177 (March 21, 2014): 1370–72, doi:10.1126/science.1249168.

53 **humans have a greater capacity:** Merlin Sheldrake, *Entangled Life: How Fungi Make Our Worlds, Change Our Minds & Shape Our Futures* (New York: Random House, 2020), 27.

54 **A public database:** Albert-László Barabási, Giulia Menichetti, and Joseph Loscalzo, "The Unmapped Chemical Complexity of Our Diet," *Nature Food* 1 (2020): 33–37, doi.org/10.1038/s43016-019-0005-1.

58 **we will live in a world:** Adrienne Rich, "Natural Resources," in *The Dream of a Common Language: Poems 1974–1977* (New York: W. W. Norton, 1978).

SIX: SUGAR SALAD

61 **twenty-five million Americans have asthma:** "Most Recent National Asthma Data," U.S. Centers for Disease Control and Prevention, last reviewed May 10, 2023, cdc.gov/asthma/most_recent_national_asthma _data.htm.

62 **Ninety percent of the vegetables:** "Potatoes and Tomatoes Are America's Top Vegetable Choices," Economic Research Service, U.S. Department of Agriculture, 2015, last updated June 5, 2018, ers.usda .gov/data-products/chart-gallery/gallery/chart-detail/?chartId=89173.

62 **foods "have been assembled":** Chris van Tulleken, *Ultra-Processed People: Why We Can't Stop Eating Food That Isn't Food* (New York: W. W. Norton), 326, Kindle.

63 **for two million years:** John Mohawk, "Clear Thinking: A Positive Solitary View of Nature," in *Original Instructions: Indigenous Teachings for a Sustainable Future*, ed. Melissa K. Nelson (Rochester, VT: Bear & Company, 2008), 48, Kindle.

63 **Seventy-five percent of young people:** Adrian Wooldridge, "A Sick America Can't Compete with China," *Bloomberg*, February 28, 2023, bloomberg.com/opinion/articles/2023-02-28/a-sick-america -can-t-compete-with-china.

63 **Forty-two percent of adults:** Nicholas Kristof, "How Do We Fix the Scandal That Is American Health Care?," *The New York Times*, August 16, 2023, nytimes.com/2023/08/16/opinion/health-care-life-expec tancy-poverty.html.

63 **we are eating exploitation:** Van Tulleken, *Ultra-Processed People*, 107, Kindle.

64 **McDonald's ditched the salad:** Colby Hall, "The Surprising Reason McDonald's Ditched This Menu Item," Eat This, Not That!, June 23, 2020, eatthis.com/mcdonalds-ditched-salads.

64 **a flow of carbon:** Eric Roston, *The Carbon Age: How Life's Core Element Has Become Civilization's Greatest Threat* (New York: Walker, 2008), 26.

65 **Our diet has changed more:** Dan Saladino, *Eating to Extinction: The World's Rarest Foods and Why We Need to Save Them* (New York: Farrar, Straus and Giroux, 2022), 2.

65 **our food became "foodlike":** Michael Pollan, *In Defense of Food: An Eater's Manifesto* (New York: Penguin, 2008), 1, Kindle.

65 **describes how humans inflict suffering:** Roston, *The Carbon Age*, 227.

65 **spends more than $5 billion:** *Fast Food Facts 2021: Fast Food Advertising, Billions in Spending, Continued High Exposure by Youth* (Hartford: University of Connecticut Rudd Center for Food Policy & Obesity, June 2021), media.ruddcenter.uconn.edu/PDFs/FACTS2021.pdf.

66 **drinks 487 cans of Coca-Cola:** Daphne Miller, *The Jungle Effect: The Healthiest Diets from around the World—Why They Work and How to Make Them Work for You* (New York: Harper Collins, 2008), 15, Kindle.

66 **One in six Mexicans:** "Diabetes Prevalence (% of population ages 20 to 79)—Mexico," World Bank Group, data.worldbank.org/indicator /SH.STA.DIAB.ZS?locations=MX.

66 **number-one cause of death:** Jason Beaubien, "How Diabetes Got to Be the No. 1 Killer in Mexico," NPR, April 5, 2017, npr.org/sections /goatsandsoda/2017/04/05/522038318/how-diabetes-got-to-be -the-no-1-killer-in-mexico.

66 **the country was flooded:** Laura Carlsen, "Under Nafta, Mexico Suffered, and the United States Felt Its Pain," *The New York Times*, November 24, 2013, nytimes.com/roomfordebate/2013/11/24/what -weve-learned-from-nafta/under-nafta-mexico-suffered-and-the -united-states-felt-its-pain.

66 **medical education focused on pathology:** Weston A. Price, *Nutrition and Physical Degeneration* (n.p.: Price-Pottenger Nutrition Foundation, 1939).

67 **food was never sold:** Melissa K. Nelson, ed., *Original Instructions: Indigenous Teachings for a Sustainable Future* (Rochester, VT: Bear & Company), 174, Kindle.

68 **Mohawk remembers at least twenty:** Nelson, ed., *Original Instructions*, 174, Kindle.

70 **have chosen to ban American:** Wayne Pacelle, "Banned in 160 Nations, Why Is Ractopamine in U.S. Pork?," Live Science, July 26, 2014, livescience.com/47032-time-for-us-to-ban-ractopamine.html.

70 **they contain ingredients considered toxic:** Christina Xenos, "Common US Foods That Are Banned in Other Countries," *Chicago Tribune*, November 3, 2021, chicagotribune.com/2021/11/03/common-us-foods -that-are-banned-in-other-countries.

70 **"Nobody has to tell a wild plant":** Fred Provenza, *Nourishment: What Animals Can Teach Us about Rediscovering Our Nutritional Wisdom* (White River Junction, VT: Chelsea Green, 2018), 7, Kindle.

SEVEN: BUCKY AND BING

71 **Spaceship Earth was a metaphor:** The idea of Spaceship Earth was first used by Henry George in his 1879 book entitled *Progress and Poverty*: "It is a well-provisioned ship, this on which we sail through space."

74 **The discovery occurred:** "Richard E. Smalley, Robert F. Curl, and Harold W. Kroto," Science History Institute, sciencehistory.org /historical-profile/richard-smalley-robert-curl-harold-kroto.

75 **The 1985 discovery riveted chemists:** Judah Ginsberg, *The Discovery of Fullerenes* (Washington, DC: American Chemical Society, October 11, 2010), acs.org/content/dam/acsorg/education/whatischemistry/land marks/fullerenes/discovery-of-fullerenes-commemorative-booklet.pdf.

75 **fullerenes set off a cascade:** Li Xiao et al., "The Water-Soluble Fullerene Derivative 'Radical Sponge®' Exerts Cytoprotective Action Against UVA Irradiation but Not Visible-Light-Catalyzed Cytotoxicity in Human Skin Keratinocytes," *Bioorganic & Medicinal Chemistry Letters* 16, no. 6 (March 15, 2006): 1590–95, doi.org/10.1016/j .bmcl.2005.12.011.

75 **structure does not dissolve:** Sergey Emelyantsev et al., "Biological Effects of C60 Fullerene Revealed with Bacterial Biosensor—Toxic or Rather Antioxidant?," *Biosensors* 9, no. 2 (2019): 81, doi:10.3390 /bios9020081.

75 **used for gene delivery:** Rania Bakry et al., "Medicinal Applications of Fullerenes," *International Journal of Nanomedicine* 2, no. 4 (2007): 639–49, pubmed.ncbi.nlm.nih.gov/18203430.

76 **doubling of the lifespan:** Tarek Baati et al., "The Prolongation of the Lifespan of Rats by Repeated Oral Administration of [60] Fullerene," *Biomaterials* 33, no. 19 (2012): 4936–46, doi:10.1016/j.biomaterials.2012.03.036.

76 **research revealed variations of fullerenes:** Ayrat Khamatgalimov et al., "Fullerenes C100 and C108: New Substructures of Higher Fullerenes," *Structural Chemistry* 32 (2021): 2283–90, doi.org/10.1007/s11224-021-01803-0.

77 **Nanotubes slough off:** Aasgeir Helland et al., "Reviewing the Environmental and Human Health Knowledge Base of Carbon Nanotubes," *Environmental Health Perspectives* 115, no. 8 (August 2007): 1125–31, doi:10.1289/ehp.9652.

77 **insoluble in water:** Rasel Das, Bey Fen Leo, and Finbarr Murphy, "The Toxic Truth about Carbon Nanotubes in Water Purification: A Perspective View," *Nanoscale Research Letters* 13, no. 183 (2018), doi.org/10.1186/s11671-018-2589-z.

77 **inhalation and exposure:** Sudjit Luanpitpong, Liying Wang, and Yon Rojanasakul, "The Effects of Carbon Nanotubes on Lung and Dermal Cellular Behaviors," *Nanomedicine* 9, no. 6 (May 2014): 895–912, doi:10.2217/nnm.14.42.

78 **The bullishness of science:** Karen F. Schmidt, *Nanofrontiers: Visions for the Future of Nanotechnologies (PEN 6)* (Washington, DC: Woodrow Wilson International Center for Scholars, Project on Emerging Nanotechnologies, March 2007), nanowerk.com/nanotechnology/reports/reportpdf/report81.pdf.

79 **found in most people alive:** Amy Westervelt, "Phthalates Are Everywhere, and the Health Risks Are Worrying. How Bad Are They Really?," *The Guardian*, February 10, 2015, theguardian.com/lifeandstyle/2015/feb/10/phthalates-plastics-chemicals-research-analysis.

79 **government regulation cannot keep up:** Ravi Naidu et al., "Chemical Pollution: A Growing Peril and Potential Catastrophic Risk to Humanity," *Environment International* 156 (2021), doi.org/10.1016/j.envint.2021.106616.

80 **uncertainties and risks:** Das, Leo, and Murphy, "The Toxic Truth about Carbon Nanotubes in Water Purification."

83 **Steel is responsible:** "What Is the Carbon Footprint of Steel?," Sustainable Ships, sustainable-ships.org/stories/2022/carbon-footprint -steel.

83 **replacing 1.7 billion tons:** James Hall, "Cleaning Up the Steel Industry: Reducing CO2 Emissions with CCUS," Carbon Clean, January 28, 2021, carbonclean.com/blog/steel-co2-emissions.

EIGHT: GREEN BEINGS

85 **"Nature needs no home":** David George Haskell, *The Songs of Trees: Stories from Nature's Great Connectors* (New York: Viking, 2017), 179, Kindle.

85 **The Mediterranean prickly oak:** Richard Mabey, *The Cabaret of Plants: Forty Thousand Years of Plant Life and the Human Imagination* (New York: W. W. Norton, 2015), 90.

86 **A seagrass colony:** Sophie Arnaud-Haond et al., "Implications of Extreme Life Span in Clonal Organisms: Millenary Clones in Meadows of the Threatened Seagrass *Posidonia oceanica*," *PLoS ONE* 7, no. 2 (2012): e30454, doi.org/10.1371/journal.pone.0030454.

86 **Plants are a measure:** Brian J. Enquist et al., "The Commonness of Rarity: Global and Future Distribution of Rarity across Land Plants," *Science Advances* 5, no. 11 (November 2019), doi:10.1126/sciadv.aaz04. Researchers did not think 35.6 percent of land species should be considered "exceedingly rare" because each was recorded less than five times over ten years and the twenty thousand million observations.

86 **Yet, grasslands, forests:** Yinon M. Bar-On, Rob Phillips, and Ron Milo, "The Biomass Distribution on Earth," *Proceedings of the National Academy of Sciences* 115, no. 25 (April 13, 2018): 6506–11, doi.org /10.1073/pnas.1711842115; "Hura crepitans (Sandbox Tree)," BioNET-EAFRINET, keys.lucidcentral.org/keys/v3/eafrinet/weeds/key/weeds /Media/Html/Hura_crepitans_(Sandbox_Tree).htm.

86 **plants developed twenty senses:** Stefano Mancuso and Alessandra Viola, *Brilliant Green: The Surprising History and Science of Plant Intelligence* (Washington, DC: Island Press, 2015); Mabey, *The Cabaret of Plants*, 3.

87 **resulting in new vocabularies:** Michael Allaby, *Oxford Dictionary of Plant Sciences*, 3rd ed., online (Oxford, UK: Oxford University Press, 2013), doi:10.1093/acref/9780199600571.001.0001.

88 **Burbank would plant thousands of seedlings:** Lynne Collins, "Luther Burbank: A Bibliographical Sketch," Luther Burbank Home and Gardens, February 1984, updated 1992, lutherburbank.org/wp-content/uploads/2023/05/Luther-Burbank-A-Biographical-Sketch.pdf.

88 **Monsanto developed glyphosate-resistant corn:** Marcus Storm, "First Map Shows Global Hotspots of Glyphosate Contamination," Sydney Institute of Agriculture, University of Sydney, March 19, 2020, sydney.edu.au/news-opinion/news/2020/03/19/glyphosate-contamination-global-hotspots-in-world-first-map.html.

89 **Today, glyphosate is the world's:** Christina Gillezeau, et al., "The Evidence of Human Exposure to Glyphosate: A Review," *Environmental Health* 18, no. 1 (2019): 2, doi:10.1186/s12940-018-0435-5.

89 **processing facilities dry:** "Pellet Mill List," *Biomass*, biomassmagazine.com/plants/list/pellet-mill.

90 **the astonishing proposition:** Gabriel Popkin, "There's a Booming Business in America's Forests. Some Aren't Happy about It," *The New York Times*, April 19, 2021, nytimes.com/2021/04/19/climate/wood-pellet-industry-climate.html.

90 **plants "have largely been reduced":** Mabey, *The Cabaret of Plants*, 4.

90 **Because they are immobile:** Mancuso and Viola, *Brilliant Green*.

91 **plants emit volatile compounds:** Ted C. J. Turlings et al., "An Elicitor in Caterpillar Oral Secretions That Induces Corn Seedlings to Emit Chemical Signals Attractive to Parasitic Wasps," *Journal of Chemical Ecology* 19 (1993): 411–25, doi.org/10.1007/BF00994314.

92 **Botanist Howard Dittmer:** Howard J. Dittmer, "A Quantitative Study of the Roots and Root Hairs of a Winter Rye Plant (Secale cereale)," *American Journal of Botany* 24, no. 7 (1937): 417–20, jstor.org/stable/2436424.

93 **emit sounds and electrically generated clicks:** František Baluška et al., "The 'Root-Brain' Hypothesis of Charles and Francis Darwin:

Revival after More Than 125 Years," *Plant Signaling & Behavior* 4, no. 12 (2009): 1121–27, doi:10.4161/psb.4.12.10574.

93 **If Mancuso is correct:** Killian Fox, "Botanist Stefano Mancuso: 'You Can Anaesthetise All Plants. This Is Extremely Fascinating,'" *The Guardian*, April 15, 2023, theguardian.com/environment/2023 /apr/15/scientist-stefano-mancuso-you-can-anaesthetise-all-plants -this-is-extremely-fascinating-tree-stories.

93 **What was the signal?:** Walter D. Koenig, "A Brief History of Masting Research," *Philosophical Transactions of the Royal Society B* 376 (2021): 20200423, doi.org/10.1098/rstb.2020.0423.

94 **In forests of lowland Borneo:** Rhett A. Butler, "Borneo," Mongabay, last update June 29, 2020, rainforests.mongabay.com/borneo.

94 **Is it possible that trees:** Melanie Jones, Jason Hoeksema, and Justine Karst, "Where the 'Wood-Wide Web' Narrative Went Wrong," *Undark*, May 5, 2023, undark.org/2023/05/25/where-the-wood-wide-web -narrative-went-wrong.

95 **Prairie dogs employ adjectives:** Patricia Dennis, Stephen M. Shuster, and Con N. Slobodchikoff, "Dialects in the Alarm Calls of Black-Tailed Prairie Dogs (*Cynomys ludovicianus*): A Case of Cultural Diffusion?," *Behavioural Processes* 181 (2020): 104243, doi: 10.1016 /j.beproc.2020.104243.

95 **Recent research indicates:** Kate Golembiewski, "Every Elephant Has Its Own Name, Study Suggests," *The New York Times*, June 10, 2024, nytimes.com/2024/06/10/science/elephants-names-rumbles.html.

95 **Is there a language of plants?:** Monica Gagliano, *Thus Spoke the Plant: A Remarkable Journey of Groundbreaking Scientific Discoveries & Personal Encounters with Plants* (Berkeley, CA: North Atlantic Books, 2018), 69.

98 **"The language of plants":** Fred Provenza, *Nourishment: What Animals Can Teach Us about Rediscovering Our Nutritional Wisdom* (White River Junction, VT: Chelsea Green Publishing, 2018), 20, Kindle.

98 **She placed potted basil:** Gagliano, *Thus Spoke the Plant*, 34.

99 **seen in open landscapes:** Monica Gagliano et al., "Out of Sight but Not Out of Mind: Alternative Means of Communication in Plants,"

PLoS ONE 7, no. 5 (2012): e37382, doi.org/10.1371/journal.pone .0037382.

99 **acoustic signals when stressed:** Muhammad Waqas, Dominique Van Der Straeten, and Christoph-Martin Geilfus, "Plants 'Cry' for Help Through Acoustic Signals," *Trends in Plant Science* 28, no. 9 (September 2023): 984–86, doi:10.1016/j.tplants.2023.05.015.

99 **How does the root tip:** Monica Gagliano, Stefano Mancuso, and Daniel Robert, "Towards Understanding Plant Bioacoustics," *Trends in Plant Science* 17, no. 6 (June 2012): 323–25, doi:10.1016/j.tplants .2012.03.002.

100 **it is only the men:** Zoë Schlanger, *The Light Eaters: How the Unseen World of Plant Intelligence Offers a New Understanding of Life on Earth* (New York: HarperCollins, 2024), 115, Kindle.

100 **those same male botanists:** Schlanger, *The Light Eaters*, 109–10, Kindle.

101 **"Literally every thought that has ever passed":** Schlanger, *The Light Eaters*, 28, Kindle.

101 **Botanical science is discovering:** Schlanger, *The Light Eaters*, 247, Kindle.

101 **What if the entire plant:** Schlanger, *The Light Eaters*, 100, Kindle.

102 **what is more important:** Jason Daley, "Humans Make Up Just 1/10,000th of Earth's Biomass," *Smithsonian*, May 25, 2018, smithson ianmag.com/smart-news/humans-make-110000th-earths-biomass -180969141.

102 **civilization will be covered:** Bar-On, Phillips, and Milo, "The Biomass Distribution on Earth."

NINE: KINDOM

103 **"spore whose time has come":** Peter McCoy, *Radical Mycology: A Treatise on Seeing & Working with Fungi* (Portland, OR: Chthaeus Press, 2016).

104 **Fungi are the connective tissue:** Merlin Sheldrake, "Mycelial Land- scapes: A Conversation with Merlin Sheldrake and Barney Steel, Moderated by Emmanuel Vaughan-Lee," in *Emergence Magazine*, Feb-

ruary 12, 2024, produced by *Emergence Magazine*, podcast, MP3 audio, 1:06:22, emergencemagazine.org/interview/mycelial-landscapes.

104 **They were identified:** McCoy, *Radical Mycology*, 2.

105 **Mycelia are vegetative threads:** Merlin Sheldrake, *Entangled Life: How Fungi Make Our Worlds, Change Our Minds & Shape Our Futures* (New York: Random House, 2020), 46.

105 **Viable spores have been found:** McCoy, *Radical Mycology*, 11.

105 **The mycorrhizal fungi:** David Hawksworth, "Mycology, A Neglected Megascience," in *Applied Mycology*, ed. Mahendra Rai and Paul D. Bridge (Wallingford, UK: Centre for Agriculture and Bioscience International, 2009), 2.

105 **The marriage of fungi and plants:** McCoy, *Radical Mycology*, 21.

105 **At some point:** Heidi Ledford, "Billion-Year-Old Fossils Set Back Evolution of Earliest Fungi," *Nature*, May 22, 2019, doi.org/10.1038 /d41586-019-01629-1.

106 **starting with delicate zooplankton:** Ed Yong, "Blue Whales Can Eat Half a Million Calories in a Single Mouthful," *National Geographic*, December 9, 2010, nationalgeographic.com/science/article /blue-whales-can-eat-half-a-million-calories-in-a-single-mouthful.

106 **rice, maize, and wheat:** "Staple Foods: What Do People Eat?," Food and Agriculture Organization, fao.org/3/u8480e/u8480e07.htm.

107 **90 percent of the Earth's soils:** Peter McCoy, "On Fungi and the Birth of the Modern Psyche," in *For the Wild Podcast*, episode 37, July 20, 2016, produced by For the Wild, podcast, MP3 audio, 57:57, forth ewild.world/listen/peter-mccoy-on-fungi-and-the-birth-of-the -modern-psyche-part-1.

108 **strategies between plants and fungi:** Gabriel Popkin, "Soil's Microbial Market Shows the Ruthless Side of Forests," *Quanta Magazine*, August 27, 2019, quantamagazine.org/soils-microbial-market-shows -the-ruthless-side-of-forests-20190827.

109 **unique to the place they are found:** Giuliani Furci, "The Inner Lives of Fungi," in *Life Worlds*, episode 3, produced by Alexa Ferminich, August 2022, podcast, MP3 audio, 57:14, lifeworld.earth/episodes-blog /fungigiulianafurci.

109 **less than 10 percent of fungi:** Patrick Greenfield, "'Unchartered Territory': More Than 2m Fungi Species Yet to Be Discovered, Scientists Say," *The Guardian*, October 10, 2023, theguardian.com/environment /2023/oct/10/uncharted-territory-kew-scientists-say-more-than-2m -fungi-species-waiting-to-be-identified-aoe.

109 **2.2 to 3.8 million more species:** David L. Hawksworth and Robert Lücking, "Fungal Diversity Revisited: 2.2 to 3.8 Million Species," *Microbiology Spectrum* 5, no. 4 (July 2017): 79–95, doi:10.1128/micro biolspec.FUNK-0052-2016.

111 **The carbon being fixed:** Heidi-Jayne Hawkins et al., "Mycorrhizal Mycelium as a Global Carbon Pool," *Current Biology* 33, no. 11 (June 5, 2023): R560–R573, doi.org/10.1016/j.cub.2023.02.027.

111 **longer chains that can remain:** Serita D. Frey, "Mycorrhizal Fungi as Mediators of Soil Organic Matter Dynamics," *Annual Review of Ecology, Evolution, and Systematics* 50, no. 1 (2019): 237–59, doi.org /10.1146/annurev-ecolsys-110617-062331.

111 **tons of greenhouse gases migrate:** Berta Bago, Philip E. Pfeffer, and Yair Shachar-Hill, "Carbon Metabolism and Transport in Arbuscular Mycorrhizas," *Plant Physiology* 124, no. 3 (November 2000): 949–58, doi.org/10.1104/pp.124.3.949.

111 **estimated two and one half billion tons:** Martin Köchy, Roland Hiederer, and Annette Freibauer, "Global Distribution of Soil Organic Carbon—Part 1: Masses and Frequency Distributions of SOC Stocks for the Tropics, Permafrost Regions, Wetlands, and the World," *Soil* 1, no. 1 (April 16, 2015): 351–65, doi.org/10.5194/soil -1-351-2015.

112 **crucial way fungi communicate:** Michael Hathaway and Willoughby Arévalo, "How Do Fungi Communicate?," *MIT Technology Review*, April 24, 2023, technologyreview.com/2023/04/24/1071363/fungi -fungus-communication-explainer.

112 ***Voyria*, commonly known as ghostplants:** Sheldrake, *Entangled Life*, 156–58.

113 **Fungi are polyglots:** Fabien Cottier and Fritz A. Mühlschlegel, "Communication in Fungi," *International Journal of Microbiology* 2012, no. 2012 (September 26, 2011): 351832, doi:10.1155/2012/351832.

113 **Sheldrake asks whether:** Sheldrake, *Entangled Life*, 161.

113 **Bees count with numbers:** Jeremy Hance, "Uncovering the Intelligence of Insects, an Interview with Lars Chittka," Mongabay, June 29, 2010, news.mongabay.com/2010/06/uncovering-the-intelligence -of-insects-an-interview-with-lars-chittka.

113 **Clark's nutcrackers remember:** University Of New Hampshire, "Researcher Uncovering Mysteries of Memory by Studying Clever Bird," *ScienceDaily*, October 12, 2006, sciencedaily.com/releases/2006 /10/061012094818.htm; Lesley Evans Ogden, "Better Know a Bird: The Clark's Nutcracker and Its Obsessive Seed Hoarding," *Audubon*, November 8, 2016, audubon.org/news/better-know-bird-clarks-nut cracker-and-its-obsessive-seed-hoarding.

TEN: PARLANCE

117 **It is a polyglot language:** Willem Larsen with "Urban Scout" Peter Michael Bauer, "E-primitive: Rewilding the English Language," Peter Michael Bauer, February 4, 2008, petermichaelbauer.com/e-primitive -rewilding-the-english-language.

117 **All of the world's:** Zhanyun Wang et al., "Toward a Global Understanding of Chemical Pollution: A First Comprehensive Analysis of National and Regional Chemical Inventories," *Environmental Science & Technology* 54, no. 5 (2020): 2575–84, doi:10.1021/acs.est.9b06379.

117 **making industrial chemistry:** Stephen Lower, "Introduction to Chemical Nomenclature," Fraser University, LibreTexts Chemistry, chem.libretexts.org/@go/page/3606?pdf.

118 **looking for the word *nature*:** Andrew Messing, "Re: Do You Know Any Examples of Indigenous Language Having a Concept for 'Wilderness?,'" ResearchGate, 2014, in response to question asked April 27, 2014, researchgate.net/post/Do-you-know-any-examples-of -indigenous-language-having-a-concept-for-wilderness/535ce7bdd2 fd6448278b45c3/citation/download.

119 **In 1520, three caravels:** Paul Hawken, *Blessed Unrest* (New York: Penguin Books, 2007), 87.

120 **like entering a different realm:** Hawken, *Blessed Unrest*, 90.

121 **more than 700 languages:** Alex Carp, "The Endangered Languages of New York," *The New York Times Magazine*, February 22, 2024, nytimes.com/interactive/2024/02/22/magazine/endangered -languages-nyc.html.

122 **the languages spoken worldwide:** "List of Endangered Languages in the United States," Wikipedia, last edited on March 12, 2024, en.wikipedia.org/wiki/List_of_endangered_languages_in_the _United_States.

122 **represent an *ethnosphere*:** Wade Davis, "The Ethnosphere and the Academy," speech given at the conference "Indigenous Knowledges: Transforming the Academy," Pennsylvania State University, May 27, 2004.

122 **The state of the world:** Christopher Moseley, ed., "Atlas of the World's Languages in Danger," UNESCO, 2010, unesdoc.unesco .org/ark:/48223/pf0000187026.

122 **Davis considers languages:** Wade Davis, *Light at the Edge of the World: A Journey through the Realm of Vanishing Cultures* (Washington, DC: National Geographic, 2001).

123 **There is a language ecosystem:** "General Information Folio 5: Appropriate Terminology, Indigenous Australian Peoples," in *Teaching the Teachers: Indigenous Australian Studies for Primary Pre-Service Teacher Education* (Oatley, Australia: School of Teacher Education, University of New South Wales, 1996), ipswich.qld.gov.au/__data/assets/pdf_file /0008/10043/appropriate_indigenous_terminoloy.pdf.

123 **They name large pine trees:** Silas Tertius Rand, *Legends of the Micmacs* (New York: Longmans, Green, & Co., 1894).

125 **California experienced historic rainfall:** William H. Brewer, *Up and Down California in 1860–1864: The Journal of William H. Brewer* (Berkeley: University of California Press, 2003).

126 **Geologic evidence shows:** Michael Dettinger and B. Lynn Ingram, "Megastorms Could Drown Massive Portions of California," *Scientific American*, January 1, 2013, scientificamerican.com/article/mega storms-could-down-massive-portions-of-california.

127 **The white man asked:** William Least Heat-Moon, *PrairyErth: A Deep Map* (New York: Houghton Mifflin Harcourt, 1991), Kindle.

127 **One third of forestlands:** "Annual Tropical Deforestation by Agricultural Product," Our World in Data, ourworldindata.org/grapher /deforestation-by-commodity.

128 **"Conscious languages do not require":** "Tiokasin Ghosthorse, Lakota Native American on Intuitive Intelligence," conversation at Tamera, Portugal, on the sidelines of the "Defend the Sacred" conference, August 17, 2019, YouTube video, 19:22, youtube.com/watch ?v=qtQ7oJKDjRg.

ELEVEN: PAPER EYES

129 **"In the name of the bee":** *The Poems of Emily Dickinson,* edited by Thomas H. Johnson, The Belknap Press of Harvard University Press, Copyright © 1951, 1955, 1979, 1983 by the President and Fellows of Harvard College.

129 **I look back with:** Yoshinori Shichida and Matsuyama Take, "Evolution of Opsins and Phototransduction," *Philosophical Transactions of the Royal Society B* 364, no. 1531 (2009): 2881–95, doi:10.1098/rstb .2009.0051.

130 **it is sporting iridescent wings:** Ben Guarino, "There's a Huge and Hidden Migration in North America—of Dragonflies," *The Washington Post,* December 21, 2018, washingtonpost.com/science/2018 /12/21/theres-huge-hidden-migration-america-dragonflies.

130 **A remarkable example:** Leland C. Wyman and Flora Bailey, *Navaho Indian Ethnoentomology* (Albuquerque: University of New Mexico Press, 1964), from Lynne Kelly, *The Memory Code: The Secrets of Stonehenge, Easter Island and Other Ancient Monuments* (New York: Pegasus, 2017), 303, Kindle; Ralph Bulmer, "Review: Untitled," review of *Navaho Indian Ethnoentomology* by Leland C. Wyman and Flora L. Bailey, *American Anthropologist* 67, no. 6 (December 1965): 1564–66, jstor.org/stable/669185.

131 **edible butterflies being ignored:** Henry Walter Bates, *The Naturalist on the River Amazons: A Record of Adventures, Habits of Animals, Sketches of Brazilian and Indian Life, and Aspects of Nature under the Equator, during Eleven Years of Travel,* 3rd ed. (New York: Humbolt, 1873; Cambridge: Cambridge University Press, 2009).

132 **The scientific explanation:** Suriya Narayanan Murugesan et al., "Butterfly Eyespots Evolved via Cooption of Ancestral Gene-Regulatory Network That Also Patterns Antennae, Legs, and Wings," *Proceedings of the National Academy of Sciences* 119, no. 8 (February 15, 2022): e2108661119, pnas.org/doi/epdf/10.1073/pnas.2108661119.

132 **When caterpillars hatch:** Max Planck Society, "Sequestration of Plant Toxins by Monarch Butterflies Leads to Reduced Warning Signal Conspicuousness," Phys.org, January 18, 2023, phys.org/news /2023-01-sequestration-toxins-monarch-butterflies-conspicuousness .html.

132 **160,000 species of moths:** Johan Eklöf, *The Darkness Manifesto: On Light Pollution, Night Ecology, and the Ancient Rhythms That Sustain Life* (New York: Scribner, 2014), 10, Kindle.

133 **One in ten described organisms:** Max Anderson, Ellen L. Rotheray, and Fiona Mathews, "Marvellous Moths! Pollen Deposition Rate of Bramble (*Rubus futicosus* L. agg.) Is Greater at Night Than Day," *PLoS One* 18, no. 3 (March 29, 2023): e0281810, doi.org /10.1371/journal.pone.0281810; Akito Kawahara, "Opinion: Look at a Moth—and Find a Wonder That's Been Waiting All Along," *The Washington Post*, August 8, 2023, washingtonpost.com/opinions/2023 /08/08/moths-environment-disappearing-photos.

133 **Could insects be:** Matilda Gibbons et al., "Can Insects Feel Pain? A Review of the Neural and Behavioural Evidence," *Advances in Insect Physiology* 63 (2022): 155–229, doi.org/10.1016/bs.aiip.2022.10.001.

134 **According to recent studies:** Irina Mikhalevich and Russell Powell, "Minds without Spines: Evolutionarily Inclusive Animal Ethics," *Animal Sentience* 29, no. 1 (2020), doi:10.51291/2377-7478.1527.

134 **They experience pain and pleasure:** Helen Lambert, Angie Elwin, and Neil D'Cruze, "Wouldn't Hurt a Fly? A Review of Insect Cognition and Sentience in Relation to Their Use as Food and Feed," *Applied Animal Behaviour Science* 243 (2021): 105432, doi.org/10.1016/j .applanim.2021.105432.

134 **are aware and conscious:** Colin Klein and Andrew B. Barron, "Insects Have the Capacity for Subjective Experience," *Animal Sentience* 9, no. 1 (2016), doi:10.51291/2377-7478.1113.

134 **"accurately regarded as aliens"**: Lars Chittka, *The Mind of a Bee* (Princeton, NJ: Princeton University Press, 2022), Kindle.

134 **diet is contained within a flower**: Stephen L. Buchmann, *What a Bee Knows: Exploring the Thoughts, Memories, and Personalities of Bees* (Washington, DC: Island Press, 2023), 57, Kindle.

135 **Insect populations are down**: Casper A. Hallmann et al., "More Than 75 Percent Decline over 27 Years in Total Flying Insect Biomass in Protected Areas," *PLoS ONE* 12, no. 10 (2017): e0185809, doi.org/10.1371/journal.pone.0185809.

135 **Insects go away, we go**: Edward O. Wilson, in Oliver Millman, *The Insect Crisis: The Fall of the Tiny Empires That Run the World* (New York: W. W. Norton, 2022), 5, 15.

135 **The Earth would regress**: Edward O. Wilson, "The Little Things That Run the World (The Importance and Conservation of Invertebrates)," *Conservation Biology* 1, no. 4 (1987): 344–46, jstor.org/stable /2386020.

136 **integral to terrestrial ecosystems**: Mark Cocker, "Look Up, Listen, and Be Very Concerned. Birds Are Vanishing—and Their Crisis Is Our Crisis," *The Guardian*, April 17, 2023, theguardian.com/com mentisfree/2023/apr/17/birds-vanishing-crisis-40m-birds.

136 **crucial to the survival**: Kenneth V. Rosenberg et al., "Decline of the North American Avifauna," *Science* 366, no. 6461 (2019): 120–24, doi:10.1126/science.aaw1313.

137 **insect crisis poses as severe**: Pedro Cardoso et al., "Scientists' Warning to Humanity on Insect Extinctions," *Biological Conservation* 242 (2020): 108426, doi.org/10.1016/j.biocon.2020.108426.

137 **Insect collapse was notably detected**: Millman, *The Insect Crisis*, 17.

138 **the dominant chemical in applesauce**: Helena Horton, "Defra May Approve 'Devastating' Bee-Killing Pesticide, Campaigners Fear," *The Guardian*, December 7, 2021, theguardian.com/environ ment/2021/dec/07/defra-may-approve-devastating-bee-killing -pesticide-campaigners-fear.

139 **farmers are addicted to an insecticide**: Courtney Lindwall, "Neonicotinoids 101: The Effects on Humans and Bees," Natural Resources

Defense Council, May 25, 2022, nrdc.org/stories/neonicotinoids-101
-effects-humans-and-bees.

139 **that will eventually destroy farming:** Stefanie Christmann, "Climate Change Enforces to Look beyond the Plant—the Example of Pollinators," *Current Opinion in Plant Biology* 56 (2020): 162–67, doi .org/10.1016/j.pbi.2019.11.001.

139 **bird populations halved in Europe:** Isabella Tree, *Wilding: Returning Nature to Our Farm* (New York: New York Review Books, 2019), 4, Kindle.

139 **Ignorance of insect ecosystems:** Frank Dikötter, *Mao's Great Famine: The History of China's Most Devastating Catastrophe, 1958–1962* (New York: Bloomsbury, 2011).

141 **Restoring farm diversity:** Stefanie Christmann et al., "Farming with Alternative Pollinators Benefits Pollinators, Natural Enemies, and Yields, and Offers Transformative Change to Agriculture," *Scientific Reports* 11, no. 1 (September 14, 2021): 18206, doi:10.1038 /s41598-021-97695-5.

142 **stop cutting the lawn:** Hilary Howard, "To Save Monarch Butterflies, They Had to Silence the Lawn Mowers," *The New York Times*, October 14, 2023, nytimes.com/2023/10/14/nyregion/to-save-monarch -butterflies-they-had-to-silence-the-lawn-mowers.html.

TWELVE: PRIMEVAL

143 **relatives of today's dragonflies:** Ker Than, "Why Giant Bugs Once Roamed the Earth," *National Geographic*, August 9, 2011, national geographic.com/science/article/110808-ancient-insects-bugs-giants -oxygen-animals-science.

144 **since the Chicxulub meteorite:** Carolyn Y. Johnson, "An Apocalyptic Dust Plume Killed Off the Dinosaurs, Study Says," *The Washington Post*, October 30, 2023, washingtonpost.com/science/2023/10/30 /dust-killed-dinosaurs-tanis-climate.

144 **plant life perished:** Peter Brannen, *The Ends of the World: Volcanic Apocalypses, Lethal Oceans, and Our Quest to Understand Earth's Past Mass Extinctions* (New York: HarperCollins, 2017), 194, Kindle.

144 **Flowering plants with dormant seeds:** Daniel Immerwahr, "Mother Trees and Socialist Forests: Is the 'Wood-Wide Web' a Fantasy?," *The Guardian*, April 23, 2024, theguardian.com/environment/2024 /apr/23/mother-trees-and-socialist-forests-is-the-wood-wide-web -a-fantasy.

145 **We have a cognitive bias:** Sarah Kaplan, "As Many as One in Six U.S. Tree Species Is Threatened with Extinction," *The Washington Post*, August 23, 2022, washingtonpost.com/climate-environment /2022/08/23/extinct-tree-species-sequoias.

145 **One in six trees:** Jayne Dowle, "Scientists Issue Stark Warning about the Threat to US Native Trees," September 24, 2022, Gardeningetc .com, gardeningetc.com/news/us-native-trees-threat.

145 **is facing extinction:** Markus Reichstein and Nuno Carvalhais, "Aspects of Forest Biomass in the Earth System: Its Role and Major Unknowns," *Surveys in Geophysics* 40 (2019): 693–707, doi.org/10.1007 /s10712-019-09551-x.

146 **The previous interglacial:** Nathaelle Bouttes, "Warm Past Climates: Is Our Future in the Past?," National Centre for Atmospheric Science, May 27, 2020, archived from the original on August 13, 2018, web .archive.org/web/20180813004809/https://www.ncas.ac.uk/en /climate-blog/397-warm-past-climates-is-our-future-in-the-past.

148 **Hippopotami wallowed in the Thames:** Thijs van Kolfschoten, "The Eemian Mammal Fauna of Central Europe," *Netherlands Journal of Geosciences* 79, no. 2/3 (2000): 269–81, doi:10.1017/S00167746000 21752.

148 **Nature doesn't plant trees:** Jean-Francois Bastin et al., "The Global Tree Restoration Potential," *Science* 365, no. 6448 (July 5, 2019): 76–79, doi:10.1126/science.aax084.

149 **Planting trees on bare land:** Ben Rawlence, *The Treeline: The Last Forest and the Future of Life on Earth* (New York: St. Martin's, 2022), 33, Kindle.

149 **Protecting existing forests would have:** Eric Roston, "Corporate Net-Zero Goals Don't Add Up to a Net-Zero Planet," *Bloomberg*, July 27, 2022, bloomberg.com/news/articles/2022-06-27/companies -net-zero-emissions-goals-don-t-add-up.

152 **When boreal forests are harvested:** Julian Mock, Nadja Popovich, and John Schwartz, "One Thing You Can Do: Help to Preserve Forests," *The New York Times*, January 8, 2020, nytimes.com/2020/01/08/climate/nyt-climate-newsletter-forests.html.

152 **picked to pieces by mining:** Rawlence, *The Treeline*, 209–10, Kindle.

152 **destroying pine trees for toilet paper:** Ian Austin and Vjosa Isai, "Canada's Logging Industry Devours Forests Crucial to Fighting Climate Change," *The New York Times*, January 4, 2024, nytimes.com/2024/01/04/world/canada/canada-boreal-forest-logging.html.

152 **"What did it take":** Zoë Schlanger, *The Light Eaters: How the Unseen World of Plant Intelligence Offers a New Understanding of Life on Earth* (New York: HarperCollins, 2024), 244, Kindle.

152 **biological scope of primary forests:** John W. Reid and Thomas E. Lovejoy, *Ever Green: Saving Big Forests to Save the Planet* (New York: W. W. Norton, 2022), 20.

153 **a colonial way of seeing:** Yinka Ibukun and Natasha White, "Dubai Firm's Africa Ambitions Raises Carbon Colonialism Concerns," *Bloomberg,* November 29, 2023, bloomberg.com/news/articles/2023-11-29/dubai-firm-s-africa-ambitions-raises-carbon-colonialism-concerns.

153 **The tropical broadleaf forests:** "Tropical and Subtropical Moist Broadleaf Forests: Southeastern Asia: Indonesia and Malaysia," World Wildlife Fund, worldwildlife.org/ecoregions/im0102.

154 **There are towering stands:** Takuo Tamakura et al., "Tree Size in a Mature Dipterocarp Forest Stand in Sebolu, East Kalimantan, Indonesia," *Southeast Asian Studies* 23, no. 4 (1986): 452–78, kyoto-seas.org/pdf/23/4/230404.pdf.

154 **"For modern humanity to keep":** Reid and Lovejoy, *Ever Green*, 8–11.

THIRTEEN: DARK EARTH

155 **the most complex living system:** Peter McCoy, *Radical Mycology: A Treatise on Seeing & Working with Fungi* (Portland, OR: Chthaeus Press, 2016), 28.

156 **It expresses the primordial truth:** Gabriel Popkin, "Soil's Microbial Market Shows the Ruthless Side of Forests," *Quanta Magazine*, August 27, 2019, quantamagazine.org/soils-microbial-market-shows-the -ruthless-side-of-forests-20190827.

157 **They feed on pests:** Anne E. Hajek and Jørgen Eilenberg, *Natural Enemies: An Introduction to Biological Control* (Cambridge: Cambridge University Press, 2018).

157 **a farmer's best friend:** Carl H. Lindroth, "The Linnaean Species of Carabid Beetles," *Zoological Journal of the Linnaean Society* 43, no. 291 (March 1957): 325–41, doi.org/10.1111/j.1096-3642.1957.tb01556.x.

157 **"planet's greatest alchemists":** Nicole Masters, *For the Love of Soil: Strategies to Regenerate Our Food Production Systems* (New Zealand: Printable Reality, 2019), 138–142, Kindle.

158 **Darwin worked closely:** Jeremy Megraw, "The Importance of Earthworms: Darwin's Last Manuscript," New York Public Library, April 19, 2022, nypl.org/blog/2012/04/19/earthworms-darwins-last -manuscript.

158 **When it comes to soil engineers:** Olga Maria Correia Chitas Ameixa et al., "Ecosystem Services Provided by the Little Things That Run the World," in *Selected Studies in Biodiversity*, ed. Bülent Şen and Oscar Grillo (London: InTechOpen, 2018), doi:10.5772 /intechopen.74847.

159 **Dung beetles improve soil structure:** "Land Degradation Neutrality," United Nations Convention to Combat Desertification, 2014, catalogue.unccd.int/858_V2_UNCCD_BRO_.pdf.

159 **a near-mystical sense of orientation:** Marie Dacke et al., "Dung Beetles Use the Milky Way for Orientation," *Current Biology* 23, no. 4 (2013): 298–300, doi:10.1016/j.cub.2012.12.034.

160 **planet of the ants:** Edward O. Wilson, *Tales from the Ant World* (New York: Liveright, 2020), 9, Kindle.

160 **colonies are larger than Texas:** Patrick Schultheiss et al., "The Abundance, Biomass and Distribution of Ants on Earth," *Proceedings of the National Academy of Sciences* 119, no. 40 (2022): e2201550119, doi.org/10.1073/pnas.2201550119.

160 **underground nests can span:** Erik Cammeraat and Anita Risch, "The Impact of Ants on Mineral Soil Properties and Processes at Different Spatial Scales," *Journal of Applied Entomology* 132, no. 4 (May 2008): 285–94, doi.org/10.1111/j.1439-0418.2008.01281.x.

160 **nematodes join the feast:** Mark Blaxter, "Nematodes: The Worm and Its Relatives," *PLoS Biology* 9, no. 4 (April 2011): 1–9, doi.org /10.1371/journal.pbio.1001050.

162 **most farmers have never seen:** Jon Stika, *A Soil Owner's Manual: How to Restore and Maintain Soil Health* (self-pub., CreateSpace, 2016), 22, Kindle.

163 **Modern tillage techniques:** Masters, *For the Love of Soil*, 158, Kindle.

164 **One third of arable land:** Michael Fakhri, "Public Statement by the United Nations Special Rapporteur on the Right to Food, Mr. Michael Fakhri," United Nations Human Rights, Office of the High Commissioner, May 20 2022, ohchr.org/sites/default/files/2022-05 /joint-statement-wto-imf-wfp.pdf.

164 **greater than natural erosion rates:** "Global Symposium on Soil Erosion," May 15–17, 2019, Food and Agricultural Organization of the United Nations, Rome, Italy, fao.org/about/meetings/soil-erosion -symposium/en.

165 **three billion people cannot afford:** Hannah Ritchie, "Three Billion People Cannot Afford a Healthy Diet," Our World in Data, July 12, 2021, ourworldindata.org/diet-affordability.

165 **The interactions between:** Tania V. Humphrey, Dario T. Bonetta, and Daphne R. Goring, "Sentinels at the Wall: Cell Wall Receptors and Sensors," *New Phytologist* 176, no. 1 (August 2007): 7–21, doi.org /10.1111/j.1469-8137.2007.02192.x.

165 **microbes can have one hundred thousand sensors:** Isabelle Hug, et al., University of Basel, "Bacteria Have a Sense of Touch," *ScienceDaily*, October 26, 2017, sciencedaily.com/releases/2017/10/171026142320 .html.

166 **a soundtrack for the dance:** Shreya Dasgupta, "Sounds of the Soil: A New Tool for Conservation?," Mongabay, June 30, 2023, news .mongabay.com/2023/06/sounds-of-the-soil-a-new-tool-for -conservation.

166 **microphones into the soil:** Ute Eberle, "Life in the Soil Was Thought to Be Silent. What If It Isn't?," *Knowable Magazine*, February 9, 2022, knowablemagazine.org/content/article/living-world/2022/life-soil-was-thought-be-silent-what-if-it-isnt.

166 **The combined utterances:** "Fascinating Soil," Sounding Soil—a Project by BioVision, soundingsoil.ch/en/know.

166 **Scientists who research:** Marcus Maeder et al., "Sounding Soil: An Acoustic, Ecological and Artistic Investigation of Soil Life," *Soundscape: The Journal of Acoustic Ecology* 18 (2019): 5–14, wfae.net/uploads/5/9/8/4/59849633/soundscape_vol18.pdf.

167 **land doctors who heal:** "Home Grown: The Agriculture Industry," The California State University, calstate.edu/csu-system/news/Pages/where-the-jobs-are-agriculture.aspx.

168 **tallgrass prairies:** "Tallgrass Prairie and Carbon Sequestration," Tallgrass Ontario, tallgrassontario.org/wp-site/carbon-sequestration.

168 **the most fertile soil:** Masters, *For the Love of Soil*, 157, Kindle.

FOURTEEN: UNTRANSLATED WORLD

169 **"Take a journey":** Johan Eklöf, *The Darkness Manifesto: On Light Pollution, Night Ecology, and the Ancient Rhythms That Sustain Life* (New York: Scribner, 2022), 216, Kindle.

170 **They traveled several times:** Isabella Tree, *Wilding: Returning Nature to Our Farm* (New York: New York Review Books, 2019), 70, Kindle.

170 **imitate what they saw:** Tree, *Wilding*, 58, Kindle.

170 **work of biologist Frans Vera:** Tree, *Wilding*, 57, Kindle.

172 **"giving nature the space":** Tree, *Wilding*, 9, Kindle.

172 **white stork nested:** Caitlin Moran, "Why the Knepp Rewilding Project Is Truly Magical," *The Times*, April 28, 2023, thetimes.co.uk/article/why-the-knepp-rewilding-project-is-truly-magical-m68trp899.

173 **The butterfly population exploded:** Tree, *Wilding*, 168, 176, 268–69, Kindle.

173 **wildland carbon capture rates:** "The Book of Wilding: Knepp's Soil Carbon Journey," Agricarbon, June 9, 2023, agricarbon.co.uk/the -book-of-wilding-knepp-soil-carbon.

174 **biological transition of the world:** Tree, *Wilding*, 9–10. Kindle.

174 **the bison were placed:** Graeme Green, "Herd of 170 Bison Could Help Store CO2 Equivalent of 43,000 Cars, Researchers Say," *The Guardian*, May 15, 2024, theguardian.com/environment/article /2024/may/15/bison-romania-tarcu-2m-cars-carbon-dioxide -emissions-aoe.

175 **a young Indigenous guide:** Robin Wall Kimmerer, *Braiding Sweetgrass: Indigenous Wisdom, Scientific Knowledge and the Teachings of Plants* (Minneapolis: Milkweed Editions, 2013), 42, Kindle.

175 **"What the young man laments":** Siddhartha Mukherjee, *The Song of the Cell: An Exploration of Medicine and the New Human* (New York: Scribner, 2023), 362.

176 **it is the difference:** Monica Gagliano, *Thus Spoke the Plant: A Remarkable Journey of Groundbreaking Scientific Discoveries & Personal Encounters with Plants* (Berkeley, CA: North Atlantic Books, 2018).

176 **linchpin of planetary life:** Simon Mustoe, *Wildlife in the Balance: Why Animals Are Humanity's Best Hope* (Melbourne, Australia: Wildiaries, 2022), 77, Kindle.

176 **little or no contact:** Brian Tomasik, "How Many Wild Animals Are There?," Essays on Reducing Suffering, 2009, last update August 7, 2019, reducing-suffering.org/how-many-wild-animals-are-there.

176 **it is unsettling to witness:** Manuela Andreoni, "What About Nature Risk?," *The New York Times*, March 14, 2024, nytimes.com /2024/03/14/climate/what-about-nature-risk.html.

177 **people in a collapsing boat:** Pema Chodron, *The Places That Scare You: A Guide to Fearlessness in Difficult Times* (Boston: Shambhala, 2002), 21.

177 **living beings have disappeared:** Lesego Chepape, "Living Planet Index: Wildlife Populations Have Declined by 69% Since 1970," *Mail & Guardian*, October 18, 2022, mg.co.za/the-green-guardian /2022-10-18-living-planet-index-wildlife-populations-have -declined-by-69-since-1970.

178 **Their home *is* our home:** Mustoe, *Wildlife in the Balance*, 58, Kindle.

180 **a pod of orcas:** Caitlin Gibson, "The Call of Tokitae," *The Washington Post*, December 5, 2023, washingtonpost.com/lifestyle/interactive/2023/tokitae-lolita-orca.

182 **anxiety and panic course through:** Tyler Austin Harper, "The 100-Year Extinction Panic Is Back, Right on Schedule," *The New York Times*, January 26, 2024, nytimes.com/2024/01/26/opinion/polycrisis-doom-extinction-humanity.html.

182 **turned the table:** Báyò Akómoláfé, "Let's Meet at the Crossroads," commencement address to Pacifica Graduate Institute, May 29, 2021, YouTube video, 1:00:17, youtube.com/watch?v=Lh2QmobEMFg, text: pgiaa.org/alumni-resources/12044.

FIFTEEN: CONSCIOUS

184 **"Sit, be still":** Jalal al-Din, translated by A. J. Arberry, *Mystical Poems of Rumi*, University of Chicago Press: 2010, Kindle Edition, 191.

191 **teachers we need are here:** Priscilla Settee, "Indigenous Knowledge as the Basis for Our Future," in *Original Instructions: Indigenous Teachings for a Sustainable Future*, ed. Melissa K. Nelson (Rochester, VT: Bear & Company, 2008), 45–46, Kindle.

192 **"If you are afraid":** Barry Lopez, *The Rediscovery of North America* (New York: Vintage, 1992), 55–57.

192 **direct experience:** Petuuche Gilbert, "Acoma Coexistence and Continuance," in *Original Instructions*, 36, Kindle.

192 **Mend and revive:** Oren Lyons, "Listening to Natural Law," in *Original Instructions*, 23–24, Kindle.

193 **"mysterious primordial intelligence":** Stephan Harding, *Animate Earth: Science, Intuition and Gaia* (New York: Chelsea Green, 2006), 40, Kindle.

194 **"plays to an empty house":** Annie Dillard, *An American Childhood* (London: Canongate, 2016), 102.

Index

100 YEARS of PUBLISHING

———◇———

Harold K. Guinzburg and George S. Oppenheimer founded Viking in 1925 with the intention of publishing books "with some claim to permanent importance rather than ephemeral popular interest." After merging with B. W. Huebsch, a small publisher with a distinguished catalog, Viking enjoyed almost fifty years of literary and commercial success before merging with Penguin Books in 1975.

Now an imprint of Penguin Random House, Viking specializes in bringing extraordinary works of fiction and nonfiction to a vast readership. In 2025, we celebrate one hundred years of excellence in publishing. Our centennial colophon will feature the original logo for Viking, created by the renowned American illustrator Rockwell Kent: a Viking ship that evokes enterprise, adventure, and exploration, ideas that inspired the imprint's name at its founding and continue to inspire us.

———◇———

For more information on Viking's history, authors, and books, please visit penguin.com/viking.